Michael M. Dediu

Newton, Benjamin Franklin, and Gauss

A chronological and photographic documentary

DERC Publishing House

Tewksbury (Boston), Massachusetts, U. S. A.

Published and printed in the
United States of America
On the Great Seal of the United States are included:
E Pluribus Unum (Out of many, one)
Annuit Coeptis (He has approved of the undertakings)
Novus Ordo Seclorum (New order of the ages)

Library of Congress Control Number: 2018900892

Dediu, Michael M.

Newton, Benjamin Franklin, and Gauss
A chronological and photographic documentary

ISBN-13: 978-1-939757-61-6

Preface

Three of the greatest personalities of the world, who had a major impact on our civilization, are Isaac Newton, Benjamin Franklin and Carl Friedrich Gauss. All three have important contributions in physics, with Newton having laid the foundations of classical mechanics, making remarkable contributions to optics, and, together with Gottfried Wilhelm Leibniz, developing the infinitesimal calculus, Benjamin Franklin having discoveries and theories regarding electricity, inventing the lightning rod, bifocals, and the Franklin stove, founding the University of Pennsylvania, and being one of the Founding Fathers of the United States, and Gauss making significant contributions to number theory, algebra, statistics, analysis, differential geometry, geodesy, geophysics, mechanics, electrostatics, magnetic fields, astronomy, matrix theory and optics, and is referred to as the *Princeps mathematicorum* ("the foremost of mathematicians").

The fascinating lives of these famous people have also other common elements besides physics: from a time perspective, the last 20 years of Newton are in the same time period with the first 20 years of Franklin, and the last 13 years of Franklin are in the same time with the first 13 years of Gauss, Newton and Gauss are very important mathematicians, and Franklin and Gauss received the same prestigious Copley Medal from the Royal Society in London, at an interval of 85 years.

This book for the general public is focused on these influential personalities, offering, in a chronological order, a variety of relevant information not only about them, but also about numerous other personalities and important events. There are also over 120 attractive and historic photographs, and I thank my wife Sophia for her photo assistance.

The more you read, the more you'll love it!

Michael M. Dediu, Ph. D.

Tewksbury (Boston), U. S. A., 19 January 2018

Michael M. Dediu is also the author of these books (which can be found on Amazon.com):

1. Aphorisms and quotations – with examples and explanations
2. Axioms, aphorisms and quotations – with examples and explanations
3. 100 Great Personalities and their Quotations
4. Professor Petre P. Teodorescu – A Great Mathematician and Engineer
5. Professor Ioan Goia – A Dedicated Engineering Professor
6. Venice (Venezia) – a new perspective. A short presentation with photographs
7. La Serenissima (Venice) - a new photographic perspective. A short presentation with many photos
8. Grand Canal – Venice. A new photographic viewpoint. A short presentation with many photos
9. Piazza San Marco – Venice. A different photographic view. A short presentation with many photos
10. Roma (Rome) - La Città Eterna. A new photographic view. A short presentation with many photos
11. Why is Rome so Fascinating? A short presentation with many photos
12. Rome, Boston and Helsinki. A short photographic presentation
13. Rome and Tokyo – two captivating cities. A short photographic presentation
14. Beautiful Places on Earth – A new photographic presentation
15. From Niagara Falls to Mount Fuji via Rome - A novel photographic presentation
16. From the USA and Canada to Italy and Japan - A fresh photographic presentation
17. Paris – Why So Many Call This City Mon Amour - A lovely photographic presentation
18. The City of Light – Paris (La Ville-Lumière) - A kaleidoscopic photographic presentation
19. Paris (Lutetia Parisiorum) – the romance capital of the world - A kaleidoscopic photographic view
20. Paris and Tokyo – a joyful photographic presentation. With a preamble about the Universe

21. From USA to Japan via Canada – A cheerful photographic documentary
22. 200 Wonderful Places, In The Last 50 Years – A personal photographic documentary
23. Must see places in USA and Japan - A kaleidoscopic photographic documentary
24. Grandeurs of the World - A kaleidoscopic photographic documentary
25. Corneliu Leu – writer on the same wavelength as Mark Twain. An American viewpoint
26. From Berkeley to Pompeii via Rome – A kaleidoscopic photographic documentary
27. From America to Europe via Japan - A kaleidoscopic photographic documentary
28. Discover America and Japan - A photographic documentary
29. J. R. Lucas – philosopher on a creative parallel with Plato, An American viewpoint
30. From America to Switzerland via France - A photographic documentary
31. From Bretton Woods to New York via Cape Cod - A photographic documentary
32. Splendid Places on the Atlantic Coast of the U. S. A. - A photographic documentary
33. Fourteen nice Cities on three Continents - A photographic documentary
34. 17 Picturesque Cities on the World Map - A photographic documentary
35. Unforgettable Places from Four Continents including Trump buildings - A photographic documentary
36. Dediu Newsletter, Volume 1, Number 1, 6 December 2016 – Monthly news, review, comments and suggestions for a better and wiser world
37. Dediu Newsletter, Volume 1, Number 2, 6 January 2017 (available at www.derc.com).
38. Dediu Newsletter, Volume 1, Number 3, 6 February 2017 (available at www.derc.com).
39. London and Greenwich, A photographic documentary
40. Dediu Newsletter, Volume 1, Number 4, 6 March 2017 (available also at www.derc.com).

41. Dediu Newsletter, Volume 1, Number 5, 6 April 2017 (available also at www.derc.com).

42. Dediu Newsletter, Volume 1, Number 6, 6 May 2017 (available also at www.derc.com).

43. Dediu Newsletter, Volume 1, Number 7, 6 June 2017 (available also at www.derc.com).

44. London, Oxford and Cambridge, A photographic documentary

45. Dediu Newsletter, Volume 1, Number 8, 6 July 2017 (available also at www.derc.com).

46. Dediu Newsletter, Volume 1, Number 9, 6 August 2017 (available also at www.derc.com).

47. Dediu Newsletter, Volume 1, Number 10, 6 September 2017 (available also at www.derc.com).

48. Three Great Professors: President Woodrow Wilson, Historian Germán Arciniegas, Mathematician Gheorghe Vrănceanu, A chronological and photographic documentary

49. Dediu Newsletter, Volume 1, Number 11, 6 October 2017 (available also at www.derc.com).

50 Dediu Newsletter, Volume 1, Number 12, 6 November 2017 (available also at www.derc.com).

51 Dediu Newsletter, Volume 2, Number 1 (13), 6 December 2017 (available also at www.derc.com).

52 Two Great Leaders: Augustus and George Washington, A chronological and photographic documentary

53 Dediu Newsletter, Volume 2, Number 2 (14), 6 January 2018 (available also at www.derc.com).

Michael M. Dediu is the editor of these books (also on Amazon.com):

1. Sophia Dediu: The life and its torrents – Ana. In Europe around 1920
2. Proceedings of the 4th International Conference "Advanced Composite Materials Engineering" COMAT 2012
3. Adolf Shvedchikov: I am an eternal child of spring – poems in English, Italian, French, German, Spanish and Russian
4. Adolf Shvedchikov: Life's Enigma – poems in English, Italian and Russian
5. Adolf Shvedchikov: Everyone wants to be HAPPY – poems in English, Spanish and Russian
6. Adolf Shvedchikov: My Life, My Love – poems in English, Italian and Russian
7. Adolf Shvedchikov: I am the gardener of love – poems in English and Russian
8. Adolf Shvedchikov: Amaretta di Saronno – poems in English and Russian
9. Adolf Shvedchikov: A Russian Rediscovers America
10. Adolf Shvedchikov: Parade of Life - poems in English and Russian
11. Adolf Shvedchikov: Overcoming Sorrow - poems in English and Russian
12. Sophia Dediu: Sophia meets Japan
13. Corneliu Leu: Roosevelt, Churchill, Stalin and Hitler: Their surprising role in Eastern Europe in 1944
14. Proceedings of the 5th International Conference "Computational Mechanics and Virtual Engineering" COMEC 2013
15. Georgeta Simion – Potanga: Beyond Imagination: A Thought-provoking novel inspired from mid-20th century events
16. Ana Dediu: The poetry of my life in Europe and The USA
17. Ana Dediu: The Four Graces
18. Proceedings of the 5th International Conference "Advanced Composite Materials Engineering" COMAT 2014
19. Sophia Dediu: Chocolate Cook Book: Is there such a thing as too much chocolate?

20. Sorin Vlase: Mechanical Identifiability in Automotive Engineering
21. Gabriel Dima: The Evolution of the Aerostructures – Concept and Technologies
22. Proceedings of the 6th International Conference "Computational Mechanics and Virtual Engineering" COMEC 2015
23. Sophia Dediu: Cook Book 1 A-B-C Common sense cooking
24. Sophia Dediu: Dim Sum Spring Festival
25. Ana Dediu & Sophia Dediu: Europe in 1985 - A chronological and photographic documentary

Table of Contents

Chapter 1. Isaac Newton

<u>1643</u> – 4 January – (25 Dec 1642 in the Julian calendar used at that time) Isaac Newton was born in Woolsthorpe Manor, United Kingdom. His mother Hannah Newton was 20 years old, and Isaac was born three months after the death of his father, a prosperous farmer, and the baby was named the same like his father: Isaac Newton. Born prematurely, young Isaac was a small child.

<u>1646</u> – January – Isaac was 3 years old when his mother Hannah Newton, 23, remarries with the Reverend Barnabas Smith, a minister of a church in a nearby village, and left Isaac in the care of her mother (and Isaac's maternal grandmother) Margery Ayscough. Isaac's mother and stepfather had three more children, therefore for Isaac one younger half-brother, and two younger half-sisters.

UK: typical houses on the road from Crewkerne to East Lambrook (8 km north, with the Anglican Church of St James from around 1150).

1649 – 30 January – Isaac was 6 when King Charles I of England (19 Nov 1600 – 30 Jan 1649 (aged 48.2), King 1625-1649) died because of Cromwell and the Puritans.

1653 – Isaac was 10 when his mother Hannah's second husband died, and she returns to live with Isaac, bringing three younger children with her, from her second marriage.

1655 – Isaac, 12, was enrolled in the King's School, Grantham Grammar School, which taught Latin and Greek, and also taught a significant foundation of mathematics.

1657 – 23 December – Josiah Franklin, the father of Benjamin Franklin, was born at Ecton, Northamptonshire, England, the ninth child of Thomas Franklin, a blacksmith-farmer, and his first wife Jane White Franklin.

UK: the George Hotel on the road from Crewkerne to East Lambrook (8 km north, with East Lambrook Manor, from around 1450).

1658 – 3 September – Isaac was 15.66 when Oliver Cromwell (25 April 1599 – 3 Sep 1658, aged 59.4, abolished monarchy in 1649, became Lord Protector from 1653 until 1658) died.

1659 – October – Isaac, 16.75, was removed from school, and living at Woolsthorpe-by-Colsterworth, where his mother, 36, attempted to make him a farmer, like Isaac's father, but he did not want. Henry Stokes, master at the King's School, persuaded Isaac's mother to send him back to school, so that he might complete his education. Isaac became the top-ranked student, distinguishing himself mainly by building sundials, and models of windmills.

1660 – 28 May – Isaac was 17 when George I of Great Britain was born in Hanover, Germany.

Charles II (29 May 1630 – 6 Feb 1685, aged 54.3) was crowned King of England at 30.

From the northwest corner of the Tower of London (left), looking southwest to the Shard. From around 1350 for 300 years the coronation procession started here at the Tower, ending at Westminster Abbey (4 km west (right)).

The northeast side of the Bloomsbury Central Baptist Church, at 235 Shaftesbury Ave, with a rose window and two towers, 300 m south of the British Museum, 200 m northeast from St. Giles-in-the-Fields (1733, in the London Borough of Camden, Palladian (Andrea Palladio (1508-1580, very influential Italian (Venice) architect) style Anglican church, with a spire and musical services).

1661 – June - Isaac, 18.4, was admitted to Trinity College, Cambridge, on the recommendation of his uncle (brother of his mother) Reverend William Ayscough, who had studied there. He started as a subsizar—paying his way by performing valet's duties—until he was awarded a scholarship.

UK, Cambridge: from Trinity Ln, looking west to the entrance of Trinity Hall (1350, a constituent college (the 5th oldest) of the University of Cambridge).

1662 – July – Isaac was 19.5 when the Royal Society was founded.

Inside London Liverpool Street Station (1874, central London railway terminus, with an atrium ceiling, in the north-eastern corner of the City of London, in the ward of Bishopsgate), train to Cambridge (founded around 50 AD (the principal Roman site is a small fort (*castrum*) Duroliponte on Castle Hill, just northwest of the city center; it was constructed around AD 70 and converted to civilian use around 50 years later. Evidence of more widespread Roman settlement has been discovered including numerous farmsteads, and a village in the Cambridge district of Newnham. Following the Roman withdrawal from Britain around 410, the location may have been abandoned), a university city on the River Cam, area 41 km^2, population 125,000, elevation 6 m, in eastern England, 80 km northeast of London, 120 km northeast of Oxford, home to the prestigious University of Cambridge (1209), to the east there is Cambridge International Airport). University of Cambridge is formed from a variety of institutions which include 31 constituent colleges, and over 100 academic departments organized into six schools. Cambridge is the second-oldest university in the English-speaking world, and the world's third-oldest surviving university. The university grew out of an association of scholars who left the University of Oxford, after a dispute with the townspeople.

1664 – Isaac, 21, was awarded a scholarship at Trinity College, for four more years. At that time, the college's teachings were based on those of Aristotle (384 BC – 322 BC (aged 62)), whom Newton supplemented with modern philosophers such as Descartes (31 March 1596 – 11 Feb 1650 (aged 53.86)), and astronomers such as Galileo Galilei (15 Feb 1564 – 8 Jan 1642 (aged 77.89)) and Thomas Street, 43, (1621-1689 (aged 68)), through whom he learned of Kepler's (27 Dec 1571 – 15 Nov 1630 (aged 58.87) work.

1665 – Newton, 22, generalized binomial theorem, and began to develop a mathematical theory that later became calculus.

August - Newton, 22.58, received his Bachelor of Arts (BA) degree from Trinity College, University of Cambridge.

The university temporarily closed as a precaution against the Great Plague. Newton moved to his mother's home in Woolsthorpe. He conducts prism experiments, discovers spectrum of light; works out his system of "fluxions," precursor of modern calculus, and begins to consider the idea of gravity.

From Trinity Ln, looking west through the entrance of Trinity Hall, (1350, by William Baterman (c 1298-1355, Bishop of Norwich between 1344 and 1355), a constituent college (the 5th oldest) of the University of Cambridge), to the Front Court and the entrance to the west building of the Front Court. To the northeast of Trinity Hall there is the separate Trinity College (1546, founder Henry VIII (1491-1547, reign 1509-1547), motto: Virtus Vera Nobilitas).

1666 – Newton was 23 when there was a big fire in London, and an outbreak of plague.

UK, London, on Sermon Lane, 100 m south of St. Paul's Cathedral, The National Firefighters Memorial (1991, by John W. Mills (born 1933)), close to the location where the big fire of 1666 started.

1667 – April – Newton, 24, returned to Cambridge.

15 August – Abiah Foger, the mother of Benjamin Franklin, was born in Nantucket, Massachusetts, British America. Abiah was the daughter of Peter Foulger (1617 (England) – 1690 (Nantucket, Mass., British America), aged 73, a Baptist missionary, schoolmaster, miller and surveyor), and Mary Morrill Foulger (1620 – 1704, aged 84, they had 9 children). Abiah was born into a Puritan family that was among the first Pilgrims to flee to Massachusetts for religious freedom, when King Charles I of England (19 Nov 1600 – 30 Jan 1649 (aged 48), King 1625-1649) began persecuting Puritans. They sailed for Boston in 1635.

October - Newton, 24.75, was elected as a fellow of Trinity College, University of Cambridge, United Kingdom.

From Trinity Ln, looking west to the entrance of Trinity Hall (1350, a constituent college (the 5th oldest) of the University of Cambridge (1209, royal charter by King Henry III (1207-1272) in 1231, motto: Hinc lucem et pocula sacra (From here, light and sacred draughts), ranked the world's fourth best university, and the first in the UK, Sir Isaac Newton (1642-1727) was a student here, Charles Babbage (1791-1871, mathematician and father of the computer) student).

1668 – Newton, 25, received his Master of Arts (MA) degree from Trinity College. His studies had impressed the first Lucasian (Reverend and politician Henry Lucas (1610-1663, aged 53)) Professor (1663-1669) Isaac Barrow, 38, (Oct 1630 – 4 May 1677 (aged 46.58)), who became master of Trinity in 1669).

1669 – Newton, 26, succeeded Isaac Barrow, 39, and, on Barrow's recommendation, was appointed the second Lucasian Professor of Mathematics at Trinity, University of Cambridge, a position he will hold for the next thirty-three years, when he will be 59. He rights *De analysi per aequationes numero terminorum infinitas,* which will be published in 1711.

UK, Cambridge, a bas-relief on the eastern wall of the western building of the Old Court (1451) of Queens' College (1448), University of Cambridge, 60 m east of the Mathematical Bridge (1749).

From Trinity Lane looking southeast to the west façade with the entrance to the Old Schools (1441, University Offices, the administrative center of the university, surrounded to the north by Gonville and Caius College (1348), to the east by the University of Cambridge Senate House (1722, where degree ceremonies are held, on King's Parade), to the south by the King's College Chapel (1446), and to the west by Trinity Hall (1350) and Clare College (1326)).

1671 – Newton, 28, publishes *Method of Fluxions*.

1672 – 11 January - Newton, one week after his 29th birthday, was elected Fellow of the Royal Society (FRS), 9.5 years after its foundation.

February - Newton's paper on optics, and his prism experiments are sent to the Society.

Newton works on the mathematics of gravitation in his home in Cambridge.

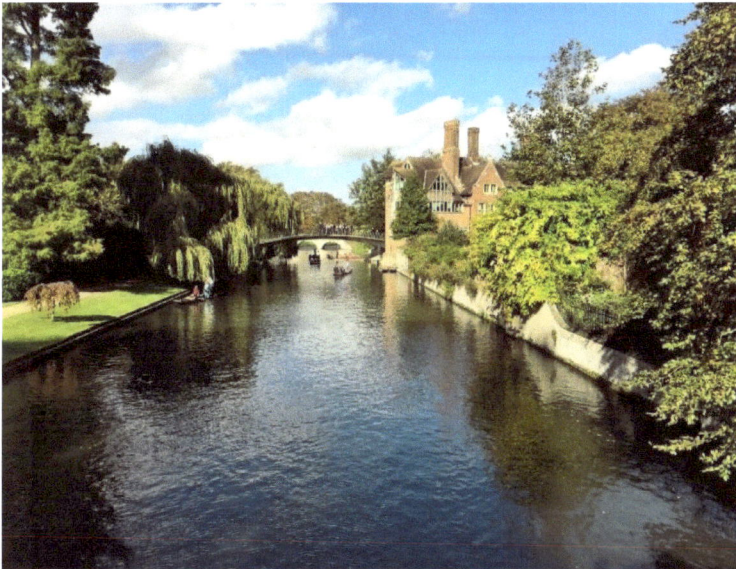

From Clare Bridge (1640, 1969) over River Cam, looking north to the Garret Hostel Bridge (center back), Punting Cambridge (center right), Trinity Hall Garden (right), Clare Fellows Garden (left).

UK, a beautiful house on Trinity Lane in Cambridge (Duroliponte in AD 70, during the Roman Empire), built in AD 1883, near the University of Cambridge (1209, includes the Cavendish Laboratory, King's College Chapel, and the Cambridge University Library). The Cambridge Biomedical Campus is one of the largest biomedical research clusters in the world. Parker's Piece, near the center, hosted the first ever game of Association football (1863).

1674 – Newton is 31 when Robert Hooke, 39, (28 July 1635 – 3 March 1703, aged 67.7, philosopher, polymath, Oxford) writes a book in which he suggests existence of "attractive powers," similar to gravity.

UK, Oxford: a Public Library close to Christ Church College (1546) and Merton College (1264, named after Walter de Merton).

UK, a splendid house in Cambridge, near King's College Chapel, which was begun in 1446 by King Henry VI (1421-1471). The chapel was built in phases by a succession of kings of England from 1446 to 1515, its history entangled with the Wars of the Roses (1455-1485), and completed during the reign of King Henry VIII (1491-1547). The Chapel building would become synonymous with Cambridge, and currently is used in the logo for the City Council.

1675 – Newton, 32, not having his views in conformity to the Church of England, received a special permission from King Charles II, now 45, King for 15 years.

At the north turn to right of Trinity Lane, a Trinity College (1546) gate, with the west part of the southern building (center right) of the Trinity College Great Court (the largest enclosed court in Europe).

UK, Oxford, on Merton Street an entrance to Corpus Christy College (1517, founder Richard Foxe, the Bishop of Winchester, 12th oldest college in Oxford, 249 undergraduates, 94 postgraduates), situated between Merton College (1264, founded by Walter de Merton (1205-1277), Lord Chancellor to Henry III (1207-1272, reign 1216-1272) and later to Edward I (1239-1307, reign 1272-1307), and Catholic Bishop of Rochester (1274-1277); Merton College Library (1373) is the oldest functioning library in the world) and Oriel College (1326).

1677 – Josiah Franklin, 19.5, married his first wife, Anne Child, and they had three children: Elizabeth, Samuel, and Hannah.

1679 – Newton was 36 when his mother Hannah Newton (1623 -1679) passed away at 56.

1682 – Josiah Franklin, 24.5, immigrated to the American colonies, in Boston, and he took up the trade of tallow chandler and soap boiler. He had other 4 children with Anne, before she died.

USA, Boston: a view of the north part of Boston, from Cambridge, over Charles River Basin. The tall building is Hancock Place (241 m) in Copley Square. To the right is The Westin Copley Place Boston Hotel, and then Boston Marriott Copley Place Hotel. To the left are Trinity Church and hotels. Storrow Drive is by Charles River.

1683 – 9 November – Newton was 40.85 when George II of Great Britain was born in Herrenhausen, Hanover, Germany.

1684 – January – Newton was 41 when, in Oxford, Hooke, 49, discusses the principle of inverse squares (from Johannes Kepler (27 Dec 1571 – 15 Nov 1630, aged 58.88, German mathematician and astronomer) with Christopher Wren, 52, (20 Oct 1632 – 8 March 1723, aged 90.4, mathematician, anatomist, astronomer, and architect, Oxford), and Edmond Halley, 27.17, (8 Nov 1656 – 14 Jan 1742, aged 85.2, mathematician, astronomer (Halley's Comet) and physicist, Oxford).

August – Halley, 27.7 goes to visit Newton, 41.6, in Cambridge, where they discuss the principle of inverse squares (Kepler), and its relationship with planetary orbits.

November – Newton. 41.8, completes his calculations on gravity, and shares them with Halley, 28, who urges him to publish.

Newton publishes *De motu corporum in gyrum*.

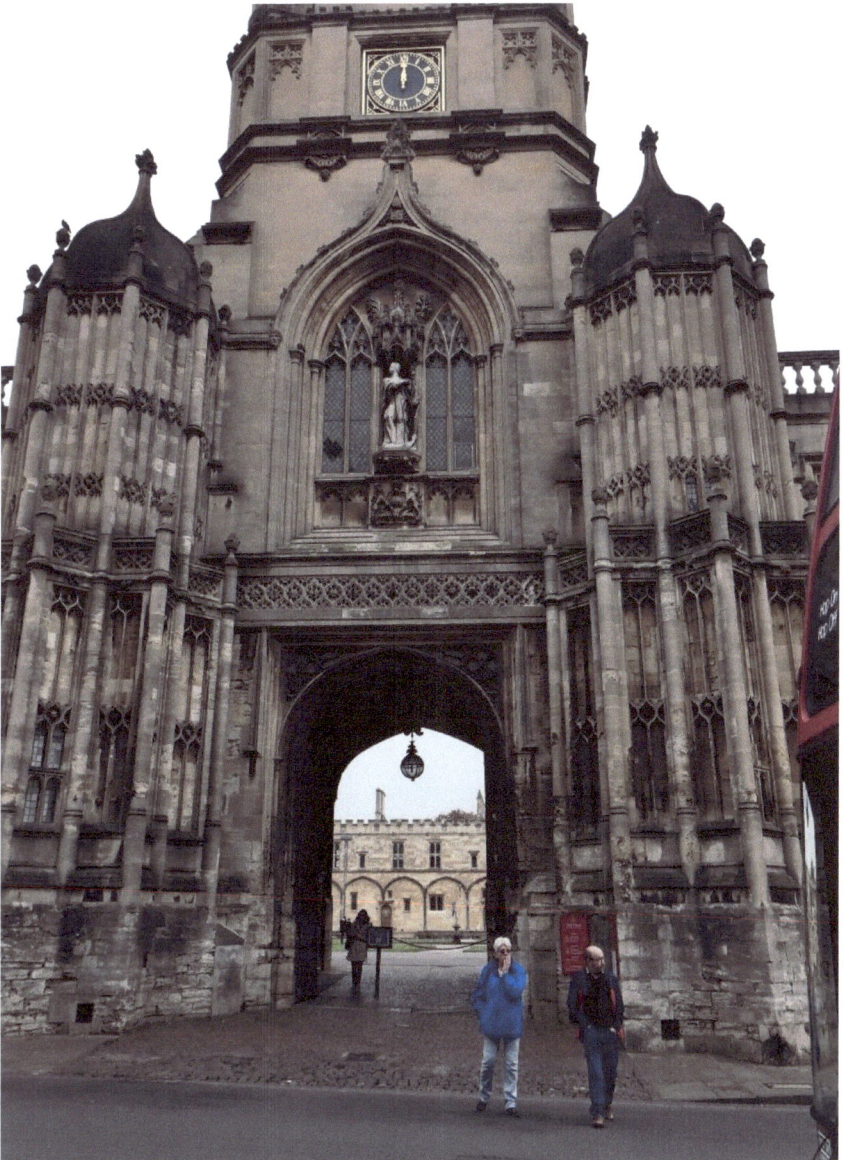

Oxford: on St Aldate's St, 200 m north of Broad Walk, 80 m east of Pembroke College, Tom Tower (1682, bell (Great Tom, rung 101 times (every 12 seconds, it takes 20 minutes) at 9 PM every night) tower) over Tom Gate, the main entrance to the majestic Christ Church College (1546, 431 undergraduates, 250 postgraduates, the second wealthiest Oxford college (after St John's), produced 13 British prime ministers), leads into its grand Tom Quad (inside).

1685 – 6 February – Newton was 42 when King Charles II died, and his brother becomes King James II (14 Oct 1633 – 16 Sep 1701, aged 67.9, King 1685 – 1688, the last Roman Catholic monarch of England, Scotland and Ireland).

February - Newton sends a brief treatise, Propositiones de Motu, to the Royal Society, outlining his findings.

1686 – April – Newton, 43, presents the first book of the Principia to the Royal Society.

UK, Cambridge: on Trinity Lane, looking east to the southern gate (center) and building of Trinity College (1546, left), and the northwestern building of Gonville and Caius College (1348, 1557, right).

UK, Cambridge, from Trinity Lane looking south to the west part of the northern façade and entrance of King's College Chapel (1446, center back, the College was founded in 1441, and the Old Schools was part of King's College), the east gate of Clare College (1326, as University Hall, making it the second-oldest college of the University, after Peterhouse (1284)) and its Chapel (1763, center right), and the Old Schools (1441, University Offices, left).

UK, London, from the northwest corner of Trafalgar Square looking northeast to part of the façade of The National Gallery (1824, 2,300 paintings) and the bronze sculpture (1686) of Jacobus Secundus (James II (1633-1701 (aged 67) reign 1685-1688)).

UK, London, on Spring Gardens, Nr. 66, near the Admiralty Arch (1912), south of Trafalgar Square.

1687 – 5 July – Newton is 44.5 when, with encouragement and financial help from Edmond Halley, 30.75, (8 Nov 1656 – 25 Jan 1742, aged 85.2, Halley's Comet), the complete book *Philosophiæ Naturalis Principia Mathematica* ("Mathematical Principles of Natural Philosophy"), was published, and laid the foundations of classical mechanics. *Principia* formulated the laws of motion and universal gravitation. By deriving Kepler's laws of planetary motion from his mathematical description of gravity, and using the same theory to find the trajectories of comets, the tides, the precession of the equinoxes, and other phenomena, Newton confirmed the heliocentric model of the Solar System (proposed by the mathematician Aristarchus of Samos (c 310 BC – c 230 BC, aged c 80), but a geometric mathematical model of a heliocentric system was presented by the mathematician, astronomer, and Catholic cleric Nicolaus Copernicus (19 Feb 1473 – 24 May 1543 (aged 70.26))), and demonstrated that the motion of objects on Earth and of celestial bodies are based on the same theory. Newton's theoretical prediction, that the Earth is shaped as an oblate spheroid, was later confirmed by the geodetic measurements of the mathematicians Pierre Louis Maupertuis (17 July 1698 – 27 July 1759 (aged 61)), Charles Marie de La Condamine (28 Jan 1701 – 4 Feb 1774 (aged 73)), and others.

UK, Cambridge: from the south gate of Trinity College (1546) Great Court (which is the main court), looking north to the northern part of the Great Court (the largest enclosed court in Europe).

<u>1688</u> – 11 December – Newton was 45 when King James II (had 15 children with two wives) was deposed, and flees to France.

<u>1689</u> – 13 February – Newton was 46 when William III, 38.2, (William of Orange, 4 Nov 1650 – 8 March 1702, aged 51.6, King 13 Feb 1689 – 8 March 1702), and Mary II, 26.78, (30 April 1662 – 28 Dec 1694, aged 32.6, Queen 13 Feb 1689 – 28 Dec 1694, wife of William III from 4 Nov 1677 (when she was 15.5, and he was 27 (on his birthday)), no children), take the throne.

Newton, 46, was elected as Cambridge's representative to Parliament, for one year, until 1690.

9 July - Josiah Franklin, 31.5, married his second wife, Abiah Folger, 21.9, (1667 – 1752, aged 85), in the Old South Church, Boston.

Abiah bore Josiah 10 children: John (1690), Peter (1692), Mary (1694), James (1697), Sarah (1699), Ebenezer (1701), Thomas (1703), Benjamin (1706), Lydia (1708), and Jane (1712).

Josiah had in total 17 children, with two wives.

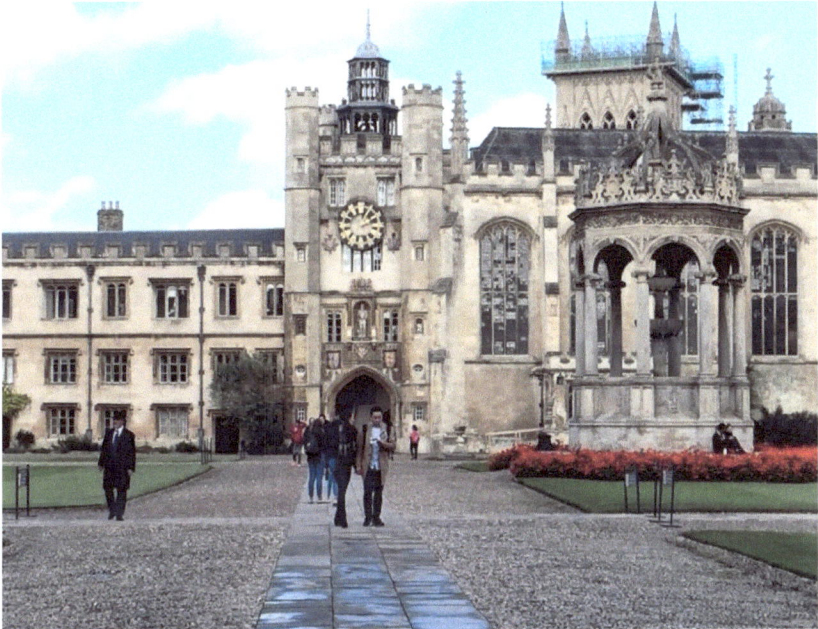

UK, Cambridge: from the south of Trinity College (1546) Great Court, looking north to the northern part of the Great Court, and the upper part of St. John's College (1511) Chapel (center right up).

1693 – September – Newton, 50.6, had depression, insomnia, and maybe suffered a nervous breakdown.

1696 – Newton, 53, was appointed Warden of the Royal Mint (for 4 years, until 1700), to oversee the implementation of a new currency, a position that he had obtained through the patronage of Charles Montagu, 1st Earl of Halifax, then Chancellor of the Exchequer. He leaves Cambridge and moves to London.

UK, Cambridge: from the center of Trinity College (1546) Great Court, looking northwest to the eastern façade of a building in the middle of the western side of the Great Court.

1700 – Newton, 57, was named Master of the Royal Mint (for 27 years, until 1727).

1701 – Newton, 58, was elected as Cambridge's representative to Parliament, for one year, until 1702. He retired from his Trinity duties.

Newton publishes *Scala graduum Caloris. Calorum Descriptiones & signa.*

UK, Cambridge: from the center of Trinity College (1546) Great Court, looking northeast to the western façade of the Main Gate (center left), and other buildings on the eastern side (right) of the Great Court, and part of the southern façade of the Trinity College Chapel (left).

1702 – 8 March – Anne, 37, (6 Feb 1665 – 1 August 1714, aged 49.5, Queen 8 March 1702 – 1 August 1714, married at 18.4, had 17 pregnancies, but no surviving children) becomes Queen.

1703 – 3 March – Newton was 60 when Robert Hooke (28 July 1635 – 3 March 1703, aged 67.7, philosopher, polymath, Oxford) died at 67.7.

Newton was elected President of the Royal Society, for 24 years, until 1727.

On St Aldate's St, 140 m north of Tom Tower, the south side and
entrance (right) of the Museum of Oxford (1975, in the former
premises of the Oxford Public Library), a history museum of the
City and University of Oxford, from prehistoric times onwards,
with original artifacts, Roman pottery, details about King Charles I
of England (1600-1649, king 1625-1649, who had Oxford as his
stronghold), Oliver Cromwell (1599-1658), etc.

1704 – Newton, 61.5, published Opticks, which included his work related to the building of the first practical reflecting telescope, the theory of color, based on the observation that a prism decomposes white light into the colors of the visible spectrum.

Newton also formulated an empirical law of cooling, made the first theoretical calculation of the speed of sound, and introduced the notion of a Newtonian fluid. In addition to his work on calculus, as a mathematician Newton contributed to the study of power series, generalized the binomial theorem to non-integer exponents, developed a method for approximating the roots of a function, and classified most of the cubic plane curves. Newton was a fellow of Trinity College. He also was interested in alchemy and biblical chronology (he was a Christian, but refused to take holy orders in the Church of England).

Gottfried Wilhelm Leibniz, 58, (1 July 1646 – 14 Nov 1716, aged 70.36, German mathematician and philosopher, who occupies a prominent place in the history of mathematics and the history of philosophy) mentions that he also has developed differential and integral calculus, independently of Newton.

1705 – Newton, 62, was knighted, and now he is Sir Isaac Newton.

UK, London: The Royal Albert Hall (1867-1871, 2004)– an Italian style concert hall on Kensington Gore, on the northern edge of South Kensington, capacity 5,272 seats, 41 m height, named after Prince Consort Albert (1819 (in Germany)-1861), husband (1840-1861) of Queen Victoria (1819-1901, Queen 1837-1901, had 9 children), Chancellor of the University of Cambridge from 1847. In July 1871, French organist and composer Camille Saint-Saëns (1835-1921) performed *Church Scene* from the Faust by Charles Gounod (1818-1893).

Chapter 2. Benjamin Franklin

1706 – 17 January – Newton was 63 years and 13 days old when Benjamin Franklin was born on Milk Street in Boston, Massachusetts Bay, English America. His father Josiah was 48, and his mother Abiah was 39. Benjamin was their youngest son.

USA, Boston: a view of the north-east part of Boston, from Cambridge, over Charles River Basin. Federal Reserve Bank Building (187 m, left), and other tall buildings in the financial district.

1707 – Newton, 64, publishes *Arithmetica Universalis*.

1712 – Newton was 69 when a Royal Society commission, under Newton's direction, investigates the competing claims of Leibniz, 66, and Newton to having developed calculus, and decides in favor of Newton.

1713 – Newton was 70 when the second edition of the Principia was published.

UK, London: from a bus near an Omega shop on Oxford Street at North Audley Street (to the right), building with columns (left), balloons (up).

From the Bow Street, the northeast façade of the Royal Opera House at Covent Garden (1732, 1808, 1858, 1999, capacity 2,256). In 1734, Covent Garden presented its first ballet, Pygmalion. On 14 January 1947, the Covent Garden Opera Company gave its first performance of Carmen (1875, opera in four acts, based on a novella of the same title by Prosper Mérimée (1803-1870 (age 67))) by French composer Georges Bizet (1838-1875 (age 36)).

<u>1714</u> – 1 August - George I, 54, (28 May 1660 – 11 June 1727, aged 67, King 1 August 1714 – 11 June 1727, 2 children) becomes King of Great Britain and Ireland.

<u>1715</u> – Benjamin, 9, has his final formal year of schooling, at Boston Latin School.

USA, Boston, 3 Dec 2009, from Avenue Louis Pasteur (1822-1895, French microbiologist), Boston Public Latin School (1635, Schola Latina Bostoniensis, the oldest and the first public exam school in the U.S.).

1716 – 14 November - Newton was 73.86 when Gottfried Wilhelm Leibniz (1 July 1646 – 14 Nov 1716) dies at 70 years 4 months and 13 days.

1717 – Benjamin, 11, begins reading Plutarch, Defoe, and Cotton Mather, invents a pair of swim fins for his hands, and is briefly indentured as a cutler.

1718 – Benjamin, 12, is apprenticed to his brother James, a printer.

1720 – Benjamin, 14, moved away from home into a boarding house, and stopped attending church so he could use Sunday to study.

At a Boston town meeting, his father Josiah, 62, is chosen as a town scavenger for 1721.

USA, Newport, the western façade of Marble House, 1888-1892, 50 rooms, 14,000 m^3 of marble, 1.6 ha, William Kissam Vanderbilt (1849-1920, younger brother of Cornelius), and his wife Alva (1853-1933).

USA, Boston: a beautiful tall ship on the north-west side of the Boston Fish Pier. Bostonians welcome tall ships and their crews and cadets, from all over the world, to their harbor, on a continuous basis.

1721 – Benjamin is 15 when his brother James Franklin starts publishing *The New England Courant*.

There is a smallpox epidemic in Boston, and controversy over vaccination.

There was at that time an intellectual movement called Deism, which accepted the existence of a creator on the basis of reason, but rejected belief in a supernatural deity who interacts with humankind. Benjamin a Deist.

1723 – Benjamin, 17, takes over the publishing of *The New England Courant,* after brother James is jailed due to "contempt" charges.

September: He runs away from apprenticeship, goes to New York, and then to Philadelphia, where he gains employment as a printer. Benjamin takes lodging with John Read, whose daughter Deborah will become Franklin's wife in 1730.

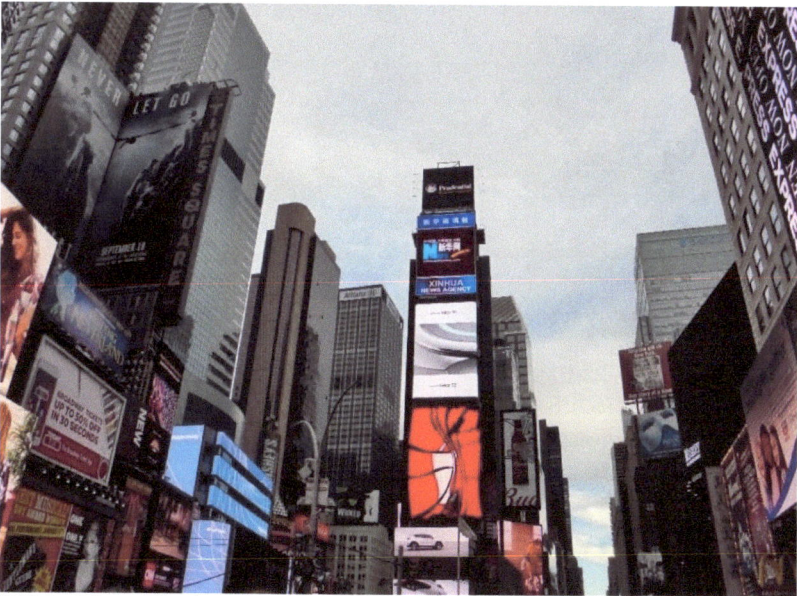

USA, New York: Times Square: 7th Ave (right straight), Broadway (center straight), W 43rd St (left and right), Marriott Marquis (right), Bertelsmann Building (left), Conde Nast Building (next), looking south.

1724 – Benjamin, 18, returns home to Boston to try to borrow money from his father, to start print shop, but is denied. Then he returns to Philadelphia, and courts Deborah Read.

Under encouragement from Pennsylvania Governor William Keith, Benjamin travels to London in order buy printing equipment. Keith's letters of credit for him never materialized, and Franklin is stranded in London. He remains in London working as a printer for Samuel Palmer.

USA, Boston, 16 April 2016, from the north side of the Boston Common, looking southeast to buildings on Boylston St (right) and Tremont St (left).

London: The Albert Memorial (1872), in southeast of Kensington Gardens, 120 m north of the Royal Albert Hall, commissioned by Queen Victoria (1819-1901, Queen 1837 (age 18)-1901 (age 82), had 9 children) in memory of her beloved husband, Prince Consort Albert (1819 (in Germany)-1861 (at 42, of typhoid)), husband (1840 (age 21)-1861 (age 42)) of Queen Victoria (same age as Albert)), Chancellor of the University of Cambridge from 1847 (age 28).

1725 – Benjamin, 19, publishes his first pamphlet: "A Dissertation upon Liberty and Necessity, Pleasure and Pain". He leaves Palmer the printer for the larger shop of John Watts, and attends theater, reads avidly, and hangs out at coffee houses.

August: in Pennsylvania, Deborah Read marries John Rogers.

1726 – Newton was 83 when the third edition of the Principia was published; all references to Leibniz have been removed.

July: Benjamin, 20, returns to Philadelphia and works for Thomas Denham, a merchant who had loaned him the money to return home. Franklin works as a bookkeeper and shopkeeper in a store which sells imported clothes and hardware.

1727 – 31 March – Sir Isaac Newton passed away at 84 years 2 months and 27 days, in Kensington, London, United Kingdom. He was buried at Westminster Abbey, London, United Kingdom. In a few words, one can say that Newton did what nobody else could have done at that time (exceptional mathematical and scientific discoveries), but he was not able to do what everybody else was doing (having a family and children).

Benjamin, 21, suffers his first pleurisy attack, leaves job with Denham, and is rehired by printer Keimer.

In this year or in 1728 Franklin has an affair with a woman that results in the birth of his illegitimate son William in 1728 or 1729.

11 June – King George I dies, aged 67, in Schloss Osnabrück, Osnabrück, Germany. His son George II, 43.6, becomes King of Great Britain and Ireland.

Franklin helps to establish the Junto, a society of young men who met together on Friday evenings for "self-improvement, study, mutual aid".

UK, London: the upper part of the western façade and entrance of Westminster Abbey (960, 1517, Anglican abbey with daily services, and all coronations since 1066, tower height 69 m).

1728 – June: Benjamin, 22, establishes a Philadelphia printing partnership with Hugh Meredith; rents a building that serves as home and printshop, and composes "Articles of Belief and Acts of Religion".

Deborah Read's husband John Rogers steals a slave and runs away from Philadelphia.

1729 – Benjamin, 23, writes a pamphlet entitled "The Nature and Necessity of a Paper Currency", and purchases *The Pennsylvania Gazette* from Samuel Keimer.

Canada, Niagara Falls: the American Falls (21-30 m drop, 290 m wide), the Bridal Veil Falls (right, 21m), after Luna Island.

1730 – Benjamin, 24, is elected the official printer for Pennsylvania.

1 September: Franklin takes a common law wife Deborah Read, 22 (14 Feb 1708 – 19 Dec 1774, aged 66.8), for 44 years.

Franklin decides to buy out his printing business partner Hugh Meredith.

Because fire destroys the southern part of Philadelphia, Franklin starts insisting for fire protection programs.

1731 – Benjamin, 25, Joins the St. Johns Freemasons Lodge.

1st July: he sketched the Library Company's articles of association. The Library Company is the first lending library in the country, and it is still private.

Franklin becomes a printing franchiser. He also prints an article in the Gazette, on the imminent passage of the "mortifying" Molasses Act, which required to pay a tax on imported molasses.

USA, New York: on 5th Ave, the southeast façade of the New York Public Library (1902).

1732 – May: Franklin, 26, started printing America's first German-language newspaper, Philadelphische Zeitung, which soon failed.

20 October: Benjamin becomes father – his son Francis Foger was born.

28 December: Franklin publishes the first edition of "Poor Richard's Almanack".

1733 – Franklin is 27 when his son Francis Folger Franklin is baptized at the Anglican Christ Church.

1734 – Franklin, 28, is elected Grand Master of the Grand Masonic Lodge of Masons of Pennsylvania. He buys property on Philadelphia's Market Street. Ultimately he will put together several lots of land on Market Street. These will house his print shop and retail space. Today, this property forms Franklin Court.

1735 – 4 Feb - Franklin is 29 when his brother James Franklin (4 Feb 1697, in Boston – 4 Feb 1735 in Newport, Rhode Island) dies on his 38th birthday. Benjamin sends his widow Ann Smith Franklin, 38.3, (2 Oct 1696 – 16 April 1763, aged 66.5, they had 5 children in 12 years of marriage (1723-1735)) 500 copies of Poor Richard for free, so she can make money by selling them.

Andrew Hamilton, a Philadelphia lawyer, defends John Peter Zenger in an important Freedom of the Press case. Hamilton will be a patron of Franklin.

USA, New York: from W 34th St, looking northeast to Seventh Ave (Fashion Ave).

1736 – Franklin, 30, is named Clerk of the Pennsylvania (PA) Assembly, and prints currency for New Jersey (NJ).

21 November: his son Francis Foger, 4 years and 1 month, passed away of smallpox.

Franklin organized the Union Fire Company, and he regularly attends meetings of the Library Company, the Masonic Lodge, the Junto, and now the Fire Company.

He prints "A Treaty of Friendship held with the Chiefs of the Six Nations at Philadelphia", and the PA State House (Independence Hall, designed by Andrew Hamilton) has its first public use.

1737 – Franklin, 31, is appointed Postmaster of Philadelphia.

USA, New York: on W 34th St, Empire State Building (center back), Snoopy second on left, looking southeast: on left a letter from the Postmaster.

1740 – Franklin, 34, is the official printer for New Jersey.

1741 – Franklin, 35, Published the first edition of "The General Magazine and Historical Chronicle," one of America's earliest magazines. It failed after six issues.

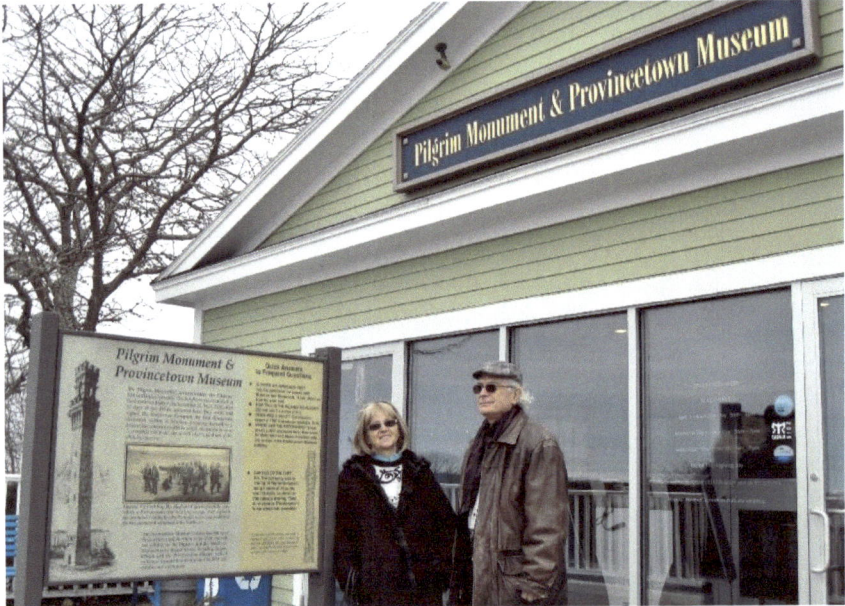

USA, Massachusetts, Cape Cod: at the entrance to Pilgrim Monument (built with granite between 1907 and 1910, commemorates the first landfall of the Pilgrims in 1620) & Provincetown Museum.

1742 – Franklin, 36, organized and publicized a project to sponsor plant collecting trips by renowned Philadelphia botanist John Bartram, 43 (23 March 1699 – 22 Sep 1777, aged 78.5).

Australia, near Sydney: in a garden in Gleenbrook (elevation 170 m. 70 km northwest of Sydney, population 5,000).

1743 – Franklin, 37, attends Archibald Spencer's, 45 (1 Jan 1698 – 13 Jan 1760, aged 62, businessman, scientist, doctor, clergyman, and lecturer) Boston lectures on natural philosophy (including the phenomenon of electricity).

He also originates "A Proposal for Promoting Useful Knowledge" (the founding document of the prototype of the American Philosophical Society).

11 Sep: his daughter Sarah (Sally) was born, latter baptized at Christ Church.

1744 – Franklin is 38 when the American Philosophical Society begins meeting

Oxford Philosopher J. R. Lucas (right, born 1932), and M. Dediu, on November 3, 2006, at the International Philosophical Conference "John Stuart Mill, 1806 – 2006".

1745 – 16 January – Franklin was one day before his 39th birthday when his father Josiah Franklin passed away in Boston, aged 87 and 24 days.

1746 – Franklin, 40, begins his numerous experiments with electricity (while the static electricity was discovered by Thales of Miletus around 600 BC, the word electricus was created by the English scientist William Gilbert around 1600, insulators and conductors were discovered by the English scientist Stephen Gray in 1720, and German physicist Ewald Georg von Kleist and Dutch scientist Pieter van Musschenbroek invented Leyden jars (the first electrical capacitors; Leyden jars store static electricity.), just in the previous year, 1745).

USA, Cambridge, 1 Feb 2010, on Massachusetts Avenue, looking northeast to the west façade of the Department of Urban Studies and Planning (left), and the main entrance to the Massachusetts Institute of Technology (MIT, 1861, center).

1747 – Franklin, 41, writes "The Plain Truth," a pamphlet arguing for better military preparedness in PA. In the pamphlet is the first political cartoon published in America.

Peter Collinson of London sends Franklin an electric tube, which captivates him: "For my own part, I never was before engaged in any study that so totally engrossed my attention, and my time, as this has lately done."

UK, near London, Greenwich, Old Royal Naval College (1712, 50 m east of Cutty Sark), now part of the University of Greenwich (1890).

1748 – Franklin, 42, becomes a soldier in the PA militia, after turning down a commission as a Colonel, citing military inexperience.

1749 – Franklin, 43, presents his vision for education in a pamphlet titled "Publick Academy of Philadelphia." His initiatives and vision would lead to the founding of the University of Pennsylvania, two years later.

USA, Boston, 16 April 2016, in the northwest corner of the Boston Public Garden, near Beacon St and Arlington St, the statue Angel of the Waters (1924 by Daniel Chester French) for George Robert White (1847 – 1922, prominent Boston businessman and philanthropist) Memorial.

USA, Cambridge, Harvard University, 23 Sep 2009, on the west side of the University Hall (1813, white granite, Colonial Revival architecture, Charles Bulfinch (1763 -1844)), in the yard of Harvard University (1636, named Harvard in 1639), in Cambridge (1630, incorporated 1636, city 1846, motto: "Literis Antiquis Novis Institutis Decora." (Distinguished (Decora) for Classical (Antiquis) Learning (Literis) and New (Novis) Institutions (Institutis)), looking northeast to the statue (1884, bronze by Daniel Chester French (1850-1931)) of the founder Reverend John Harvard (1607-1638, English minister in American colonies, (his grandfather (from the mother side) Thomas Rogers (1540-1611) was an younger associate of John Shakespeare (1531-1601), father of William Shakespeare (1564-1616)), had bequeathed to the school his entire library and half of his monetary estate).

1751 – Franklin, 45, initiated and was the first president of Academy and College of Philadelphia, which opened in this year, and later became the University of Pennsylvania.

Franklin's letters on electricity *Experiments and Observations on Electricity,* were published in London by Peter Collinson, 57, (Jan 1694 – 11 August 1768, aged 74.5, Fellow of the Royal Society, botanist, avid gardener, horticultural friend with John Bartram, and had important correspondence with Benjamin Franklin about electricity).

UK, London: on Newington Butts St., looking northwest to Metropolitan Tabernacle (1650, left), and London College of Communication.

1752 – Franklin was 46 when his mother Abiah passed away at 85.

Franklin conducts the kite experiment: he tied a key to a kite string during a thunderstorm, and proved that static electricity and lightning were the same thing.

Franklin wrote a plan for a union of the colonies for security and defense. He also helps found the Philadelphia Contributionship for Insuring of Houses from Loss Against Fire.

USA, Massachusetts, Worcester, 23 Jan 2010, Stoddard Laboratories (left), Salisbury Laboratories (center) and Fuller Laboratories (right) of the Worcester Polytechnic Institute (WPI, 1865, private research university in Worcester, 6,000 students).

1753 – Franklin, 47, received honorary degrees from Harvard and Yale.

He received the Copley Medal of the Royal Society of London for his research in electricity. The Copley Medal is the Royal Society's oldest and most prestigious award, also the world's oldest scientific prize, named for Sir Godfrey Copley, 2nd Baronet (c. 1653–1709, aged 56), a member of the Royal Society, and longtime Member of Parliament from Yorkshire.

He also was promoted to deputy postmaster-general for the British colonies in North America, having been Philadelphia postmaster for 16 years, and this enabled him to set up the first national communications network.

USA, Cambridge, 23 September 2009, on the campus of Harvard University (1636) in Cambridge, The Harry Elkins Widener (1885-1912 (died on Titanic)) Memorial Library (1915, Beaux-Arts architecture, 3.5 M of books).

UK, London, on Broad Ct looking northeast, off Bow Street to the northeast, 50 m north of the Royal Opera House at Covent Garden (1732, 1808, 1858, 1999, capacity 2,256; in 1734, Covent Garden presented its first ballet, Pygmalion), the bronze statue Young Dancer, by the Italian-born (in Mestre, near Venice, in 1921) British sculptor Enzo Plazzotta (1921-1981 (age 60)). To the right up, five red telephone boxes, at 5 Broad Ct, a tourist attraction.

1754 – Franklin, 48, proposes a plan for a colonial union, at an Albany Congress.

1757 – Franklin, 51, is sent to England as agent for Pennsylvania Assembly, Massachusetts, Georgia, and New Jersey, for 5 years.

1759 – Franklin, 53, receives honorary degree of Doctor of Laws from the University of St. Andrews, Scotland.

1760 – 25 October – Franklin is 54, in England, when King George II of Great Britain and Ireland dies in Kensington Palace, London, United Kingdom, aged 76.9, 15 days before 77. His grandson George III, 22.3, becomes King of Great Britain and Ireland.

UK, London: from Abingdon St, 50 m south of Westminster Abbey and 80 m west of Palace of Westminster, looking west to the statue of King George V (1865-1936 (aged 70.4), King 1910 – 1936 (25.7), grandson of Queen Victoria (who was granddaughter of King George III), and grandfather of the Queen Elizabeth II).

UK, London, from Charing Cross Rd, looking southeast to the northwest part of the front part of the English Anglican church St Martin in the Fields (1724, at the northeast corner of Trafalgar Square in the City of Westminster, spire height 59 m, 12 bells, tenor bell weight 1,486 kg, excavations under found a grave from about 410 AD (Roman era), in 1222 there was a church here, in 1542 Henry VIII rebuilt the church, in 1606 James I enlarged the church). It is famous for its regular lunchtime and evening concerts; Academy of St Martin-in-the-Fields performs here, and many other ensembles.

1762 – Franklin, 56, returns from England, and mapped Postal routes in the British colonies in America. He also invents glass armonica.

1764 – May - Franklin, 58.2, is elected Speaker of the Pennsylvania Assembly, until October. He also works two years to chart the Gulf Stream.

1766 – Franklin, 60, is in England to be examined in the House of Commons in support of the repeal of the Stamp Act.

UK, London: from the Abingdon St., looking northeast to the south (left, House of Commons) and west (right, House of Lords) façades of the Palace of Westminster (1016, 1870), with the Old Palace Yard, and the statue of the King Richard I of England (Coeur de Lion (the Lionheart), 1157-1199 (aged 42), King 1189-1199 (10 years), center).

1767 – Franklin is 61 when his daughter Sarah, 24, marries Richard Bache.

1768 – Franklin, 62, is named Colonial Agent for Georgia.

He was associated with the *Pennsylvania Chronicle*, a newspaper that was known for its revolutionary sentiments, and criticisms of the British policies

1769 – Franklin, 63, is named Colonial Agent for New Jersey.

He organized and was the first secretary of the American Philosophical Society, and was elected president in this year.

15 August - Napoleone di Buonaparte was born in Corsica.

1770 – Franklin, 64, is named Colonial Agent for Massachusetts.

USA, Massachusetts, Cape Cod: a monument at the southwest of Provincetown: The first landing Place of the Pilgrims on November 11, 1620, erected in 1917.

1771 – Franklin, 65, tours Ireland. He also begins writing his Autobiography.

1774 – Franklin, 68 is dressed down, before London's Privy Council, by Solicitor General Wedderburn, for leaking letters in the "Hutchinson Affair."
19 Dec: his wife of 44 years Deborah passes away of stroke, in Philadelphia, at 66.8.

1775 – Franklin, 69, is elected as a delegate of Pennsylvania (PA) to the Second Continental Congress; he serves as chairman of Pennsylvania Committee of Safety.
26 July - Franklin, 69.5, was appointed as the first Postmaster-General of the British Colonies in America.

USA, Massachusetts, Cape Cod: The Pilgrims Monument in the center of Provincetown, built with granite between 1907 and 1910, commemorates the first landfall of the Pilgrims in 1620, and the signing in Provincetown Harbor of the Mayflower Compact.

Inside (southwest) the British Museum (1753). Head of the Egyptian pharaoh Amenhotep III (1411 BC – 1353 BC (age 58), pharaoh for 33 years (1386 BC (age 25) – 1353 BC), Egyptian Amana-Hatpa meaning god Amun is Satisfied, also known as Amenhotep the Magnificent, the 9th pharaoh of the 18th Dynasty (1570 BC – 1293 BC)) wearing the red crown of Lower Egypt, Amana period at the end of the 18th Dynasty, 1360 BC, from Thebes. One of his grandsons was Tutankhamun (1342 BC – 1323 BC (age 19)).

1776 – Franklin, 70, presides over Constitutional Convention of PA.

He serves on a committee of five who draft the Declaration of Independence.

4th of July: the United States Declaration of Independence (Thomas Jefferson (33, Virginia) was the principal author, with Benjamin Franklin (70, Pennsylvania) and John Adams (40.7, Massachusetts)) was adopted by the Second Continental Congress meeting at the Pennsylvania State House (Independence Hall) in Philadelphia, which announced that the thirteen American colonies, then at war with the Kingdom of Great Britain, regarded themselves as thirteen independent sovereign states, no longer under British rule. It was signed by 56 signers.

7 November - Franklin, 70.8, leaves the position of the first Postmaster-General.

21 December: Benjamin Franklin, now 70.9 years old, arrives in Paris as one of the Commissioners of Congress to the French Court, being dispatched to France to rally its support, and he was welcomed with great enthusiasm.

France, Paris: La Monnaie de Paris (the Direction of Coins and Medals) created in 864 by Charles II (823-877, king 843-877), is the oldest French institution, which is still active. It also has a Musée de la Monnaie (1833), at 11 Quai de Conti, in the 6th arrondissement.

Chapter 3. Johann Carl Friedrich Gauss

<u>1777</u> – 30 April - Johann Carl Friedrich Gauss was born in Braunschweig (now Lower Saxony, Germany), main city of the Duchy of Brunswick-Wolfenbüttel, Holy Roman Empire. His father was a former small farmer turned urban laborer, his mother from a family of stonemasons.

Franklin, 71, meets Madame Anne Louis Brillon de Jouy, 33, (1744 – 1824 (died 8 days before 80), French musician, wife of Monsieur Brillon de Jouy), who becomes an amour.

31 July: After Lafayette offered to serve without pay, Congress, at the request of Benjamin Franklin, commissioned Lafayette a major general, at 19.8. Lafayette was married from 1774, when he was 17.

France, Paris: Galeries Lafayette (1895) a classy store on Blvd Haussmann (right), near Rue La Fayette (center-left).

1778 – Franklin, 72, is received by the King of France Louis XVI (24 years old, 1754-1793, King of the Bourbon dynasty 1774-1792). He secured the military support of France.

6 Feb: France formally recognized the United States, with the signing of the Treaty of Alliance (Franklin signed the Treaty). The situation of the American war was radically altered.

17 March: Britain declared war on France.

14 September - Franklin, 72.7, is appointed by the Continental Congress as United States Minister to France. He will serve with Arthur Lee, Silas Deane and John Adams.

France, Paris: From l'église de la Madeleine (1842) – Rue Royale, the Egyptian obelisk (circa 1250 BC) in Place de la Concord (1772), and Palais Bourbon (1728, Napoleon 1806), now for l'Assemblée Nationale.

From the northeast corner of Trafalgar Square, south of the National Gallery, looking southwest to Vice Admiral Horatio Nelson's (1758-1805 (aged 47), buried at St Paul's Cathedral) Column, and the equestrian statue of King George IV (1762-1830 (aged 68), King 1820-1830, patron of architecture, the eldest son of King George III (1738-1820 (aged 81), Reign 1760-1820 (59 years), during his reign, the American colonies created the U. S. A.)).

1779 – Franklin, 73, is appointed for two years to negotiate a peace treaty with England.

1780 – Franklin, 74, wanted to marry Anne-Catherine de Ligniville, Madame Helvétius, 58, (nicknamed "Minette", 1722 – 1800, aged 78), but she did not want.

Gauss was only 3 years old when he corrected, mentally and without mistakes, an error his father had made on paper, while calculating finances.

1782 – 28 September – Franklin, 76.7, is appointed by the Congress of the Confederation as United States Minister to Sweden.

Sweden, Malmö, from Skeppsbron looking north to the north part of the west side of the Central Station (right), sign for Trelleborg and Limhamn (to left), Goteborg and Hamnen (straight).

1783 – 3 April - Franklin, 77.2, leaves his position of United States Minister to Sweden.

3 September: Paris: The Treaty of Paris (initiated on 30 Nov 1782 by the King of France Louis XVI, and signed at the Hotel d'York (presently a private building at 56 Rue Jacob, with a plaque on the left side of the building, which recalls the historic event; this place is just 100 m southwest of Ècole National Supérieure des Beaux-Arts (1682, National Advanced School of Fine Arts))) was signed in Paris on September 3, 1783, by John Adams (48, 1735 – 1826), Benjamin Franklin (77) and John Jay (38, 1745 – 1829, Founding Father, diplomat, first Chief Justice) (for the US), and David Hartley (51, 1732 – 1813, statesman, scientific inventor, and the son of the philosopher David Harley (1705 – 1757), for the Great Britain), resulted in the peace between the Great Britain and the U. S. and, at Versailles, peace with US allies France (King Louis XVI, 29), Spain (King Charles III (67, 1716-1788, King 1759-1788)) and the Dutch Republic (Pieter van Bleiswijk (59, 1724-1790, Grand Pensionary 1772-1787)). The most important sentence of the treaty is: "His Britannic Majesty acknowledges the United States of America to be free, sovereign and independent." It was effective on 12 May 1784.

25 November: the evacuation of the British from New York City took place. Washington entered with the American army into New York City.

Franklin also invented bifocals.

From 1783, Georges Washington La Fayette (4) grew up in the Hôtel de La Fayette at 183 rue de Bourbon, Paris. Their home was the headquarters of Americans in Paris. People such as Benjamin Franklin, Mr. and Mrs. John Jay, and Mr. and Mrs. John Adams met there every Monday. They dined with the La Fayette family, as well as with the liberal nobility, such as Clermont-Tonnerre, Madame de Staël, Morellet, and Marmontel.

Through the next years, Lafayette made his house, the Hôtel de La Fayette in Paris's rue de Bourbon, the headquarters of Americans there. Benjamin Franklin, John and Sarah Jay, and John and Abigail Adams met there every Monday, and dined in company with Lafayette's family and the liberal nobility, including Clermont-Tonnerre and Madame de Staël. Lafayette continued to work on lowering trade barriers in France to American goods, and on assisting Franklin and his successor as envoy, Jefferson, in seeking treaties of amity and commerce with European nations.

France, Paris: Place de la Concorde: the north side of the Egyptian obelisk (circa 1250 BC), with hieroglyphics about the pharaoh Ramses the Great (1303 BC – 1213 BC (90 years), reign 1279 BC – 1213 BC (66 years)). The obelisk is from Luxor, rises 23 m, weights 250 t and it was placed here by the King Louis Philippe I (1773 – 1850, reign 1830 – 1848) in 1836, On the pedestal are drawn diagrams showing the techniques used for transportation. The original cap was stolen in Luxor around 550 BC, and the French Government added a gold-leafed pyramid cap in 1998.

1785 - 17 May - Franklin, 79.3, leaves his position of United States Minister to France. He is succeeded by Thomas Jefferson.

18 October – Franklin, 79.7, was elected the 6th President of Pennsylvania Executive Council.

He also invented the "long arm," an instrument for taking down books from a shelf.

Gauss, 8, was a child prodigy - a story relates that he figured out how to add up fast all the numbers from 1 to 100.

1787 - Franklin, 81, signed the United States Constitution.

USA, Washington, D.C. (1790): flags on the National Mall, with the Capitol (1793 – 1800, 88 m, center back), and the Smithsonian Institution Building (1849-1855, center right).

1788 5 November – Franklin, 82.8, ends his position of President of Pennsylvania Executive Council.

1789 - Franklin, 83, becomes president of the Society for Promoting the Abolition of Slavery.

1790 - 17 April: George Washington, 58, was informed that Benjamin Franklin passed away, of pleurisy, at 84 years and 3 months, in Philadelphia, Pennsylvania, USA. 20,000 mourners attend his funeral at Philadelphia's Christ Church Burial Ground.

His status as one of America's most influential Founding Fathers, have seen Franklin honored on coinage and the $100 bill, warships, and the names of many towns, counties, educational institutions, and corporations, as well as numerous cultural references.

Gauss was 13.

Benjamin Franklin quotes: "An investment in knowledge pays the best interest".

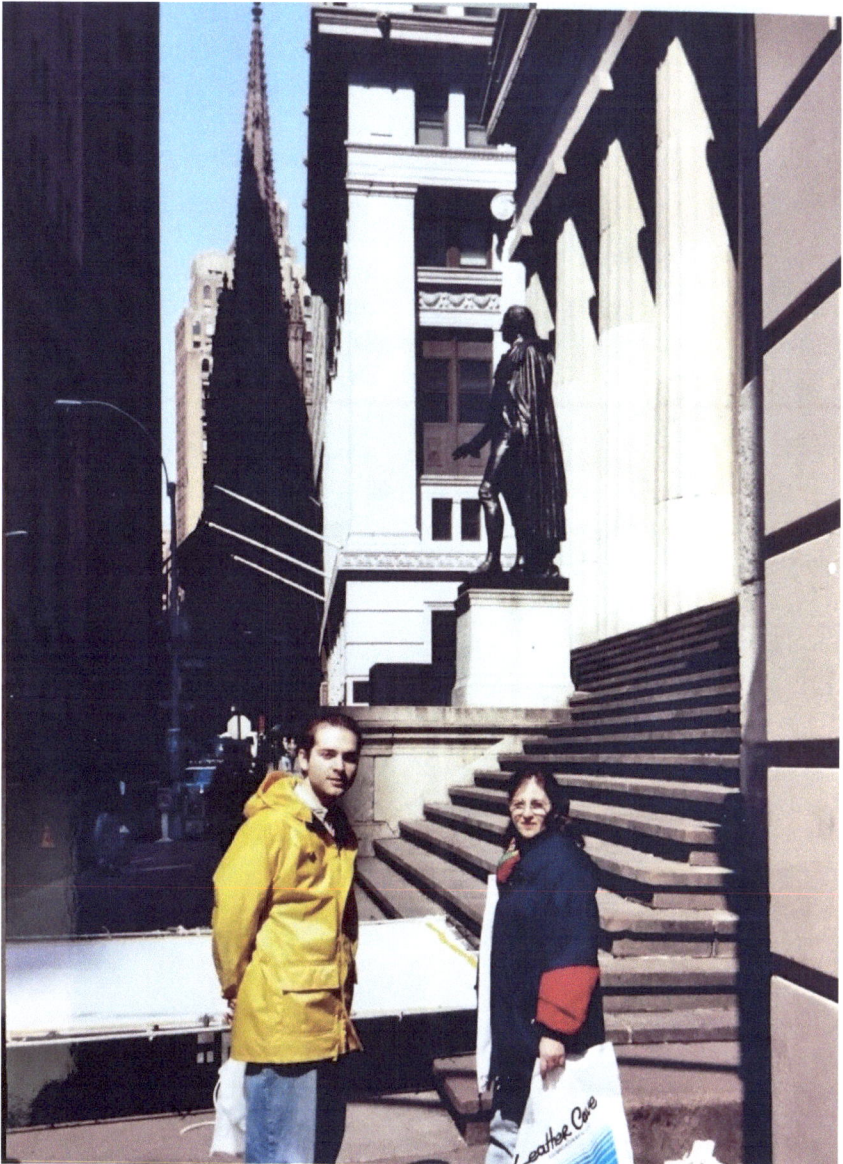

New York City, on Wall Street, looking northwest to 26 Wall Street, Federal Hall National Memorial (right, the current building opened in 1842), on April 30, 1789, George Washington (1732-1799) was inaugurated on the balcony of the previous building (the United States' first presidential inauguration) – his statue (1882) is in the front of the building (center right), and the Trinity Church (1846, 87 m height, at 75 Broadway, center left back).

1791 – Gauss, 14, was introduced to Charles William Ferdinand, 56, (9 Oct 1735 – 10 Nov 1806 (aged 71), Duke of Brunswick 26 March 1780 – 10 Nov 1806 (26 years)), who awards him an annual stipend (scholarship); Gauss would receive financial support from the Duke until 1806 (for 15 years), when he will be 29.

1792 – Gauss, 15, was sent by the Duke of Brunswick to secondary studies at the Collegium Carolinum (now Braunschweig University of Technology), for 3 years, until 1795. He reads more mathematics on his own (Newton, Leonhard Euler (15 April 1707 in Basel, Switzerland – 18 Sep 1783 in Saint Petersburg, Russia (aged 76.4), Swiss mathematician), Joseph-Louis Lagrange, now 56, (born Giuseppe Lodovico Lagrangia on 25 Jan 1736 in Torino, Italy – 10 April 1813 in Paris, France (aged 77.2), Italian mathematician) that he can begin to engage in original research.

Switzerland: from Genève to Thoiry (France), on Route de Meyrin, 2 km west from Geneva Cointrin Airport, there is this renovation of the external structural Elements of the Globe of Science and Innovation.

1795 – Gauss, 18, studies for 3 years at the recently established University of Göttingen, chosen for its good library and science orientation. He met Wolfgang Bolyai, 20, (9 Feb 1775 in Șeica Mare (now in Romania) – 20 Nov 1856 in Târgu Mureș (now in Romania), aged 81.7), a philosophy and mathematics student. Gauss independently rediscovered several important theorems.

Romania, Sibiu (25 km south of Șeica Mare, and 90 km southwest of Târgu Mureș), 11 Oct 2008, in the Small Square, looking northwest to the Liars Bridge (1859, the first footbridge in Romania to have been cast in iron, its name comes from stories and tall talk of the nearby hagglers selling fish, center), and a bride being photographed (down).

1796 – Gauss, 19, keeps a mathematical diary, for 5 years, until 1801, with 146 entries; the first entry is on the heptadecagon (he showed that a regular polygon can be constructed by compass and straightedge if the number of its sides is the product of distinct Fermat primes and a power of 2). He thinks about the distribution of prime numbers (among many other topics), and works on *Disquisitiones Artihmeticae.*

30 March – Gauss, 18.91, discovered a construction, only with a compass and straightedge, of the heptadecagon, a regular 17-sided polygon.

8 April - Gauss, 18.93, became the first to prove the quadratic reciprocity law.

31 May – At 19 and 1 month, Gauss conjectured the prime number theorem, which gives a good understanding of how the prime numbers are distributed among the integers.

10 July – At 19.19 he discovered that every positive integer is representable as a sum of at most three triangular numbers.

1 October – At 19.4 Gauss published a result on the number of solutions of polynomials with coefficients in finite fields, which 153 years later led to the Weil (André Weil: 6 May 1906 – 6 Aug 1998, aged 92 years and 3 months) conjectures (1949).

1 Feb 2010, geometrical shapes presented at MIT Mathematics Department, including octahedrons (left up, with 8 faces, 12 edges and 6 vertices; a regular octahedron has equilateral triangles for its faces, and is one of the 5 platonic solids), dodecahedrons (with 12 faces, 30 edges and 20 vertices; a regular dodecahedron has regular pentagons for its faces, and is one of the 5 platonic solids), icosahedron (with 20 faces, 30 edges and 12 vertices; all the faces are triangles; a regular icosahedron is one of the 5 platonic solids with all faces being equilateral triangles).

92

1797 – 4 March - Gauss was 19.8 when George Washington, 65, refusing a third term, retired from the presidency and all public life, and John Adams, 61.3 (30 Oct 1735 in Quincy Mass. – 4 July 1826 in Quincy Mass. (aged 90.7), a Founding Father, the first Vice President, was elected the second President of the United States (1797 – 1801).

19 Sep - in France, Lafayette, 40, returned to France from a prison in Austria, after Napoleon Bonaparte, 28, secured his release, though Lafayette refused to participate in Napoleon's government. From Hamburg, Lafayette sent a note of thanks to General Napoleon Bonaparte.

Italy, 6 April 1978, Pisa, Palazzo della Carovana (1562-1564) now for Scuola Normale Superiore (1810, by Napoleon Bonaparte (1769-1821), 460 students, 6% admission rate, best in Italy).

1798 – Fall - Gauss, 21.5, leaves University of Göttingen without a diploma, and returns to Brunswick, 90 km northeast from Göttingen. Here he was in contact with astronomers Franz Xaver von Zach, 44, (4 June 1754 – 2 Sep 1832, aged 78.2) and Heinrich Wilhelm Matthias Olbers, 40, (11 Oct 1758 – 2 March 1840, aged 81.3). Until 1807 Gauss developed an interest in mathematical and observational astronomy, and the idea of working as an astronomer, rather than a mathematics professor.

Gauss completed *Disquisitiones Arithmeticae*, published in 1801. This work was fundamental in consolidating number theory as a discipline, and has shaped the field to the present day.

Switzerland, Neuchâtel (570 km southwest of Göttingen), on Avenue du Premier Mars at Rue Coulon (right), the northwest side of the Université de Neuchâtel (1838, 4,400 students).

Germany (southwest), 1978, Oberwolfach (the district of Ortenau in Baden-Württemberg, elevation 323 m (270 m to 948 m), 465 km southwest of Brunswick, and 375 km southwest of Göttingen, in the central Schwarzwald (Black Forest) on the river Wolf, a tributary of the Kinzig.): Academician Professor Dr. Gheorghe Vranceanu (right) and Dr. Michael Dediu at the entrance to the Mathematisches Forschungsinstitut Oberwolfach (Mathematical Research Institute of Oberwolfach, founded in 1944 by the German mathematician Wilhelm Süss (1895-1958)).

1799 – Gauss, 22, submits, at the Duke's request, a doctoral dissertation to the University of Helmstedt, containing a proof of the Fundamental Theorem of Algebra: *Demonstratio nova theorematis omnem functionem algebraicam rationalem integram unius variabilis in factores reales primi vel secundi gradus resolvi posse* ("New proof of the theorem that every integral algebraic function of one variable can be resolved into real factors (i.e., polynomials) of the first or second degree"). Doctoral advisor was Johann Friedrich Pfaff, 33.5, (22 Dec 1765 – 21 April 1825, aged 59.3).

14 Dec - George Washington passed away, in his bedroom at Mount Vernon, at 67 years, 9 months and 22 days.

Later, in France, Napoleon Bonaparte, 30, held a memorial service in Paris for Washington, but Lafayette, 42, was not invited, nor was his name mentioned.

France, Lyon (43 BC), from Place de la Comédie, looking northeast to the northwest part of the Opéra (right), and of the Place Louis Pradel.

USA, Boston (1629, motto: Sicut patribus sit Deus nobis (As God was with our fathers, so may He be with us), the capital and largest city of the Commonwealth of Massachusetts, with the first US public school, Boston Latin School (1635), and first subway system (1897), international center of higher education, medicine and innovation), 20 June 2015, Boston Public Garden (1837, 9.7 ha, 1 km south of MGH, adjacent to the southwest of the Boston Common), the equestrian statue of George Washington (1732-1799), by Thomas Ball in 1869, near the Arlington Street gate.

1800 – 1 March – Gauss was 22.8 when, in France, Napoleon's (30.5 years old) plebiscite took place, after which he restored La Fayette's (42.5) citizenship, and removed his family's names from the émigrés list.

1801 – Gauss, 24, ended his mathematical diary, and publishes (in Leipzig) *Disquisitiones Artihmeticae,* the foundational work of algebraic number theory. There are seven sections: the first three are a general introduction to the arithmetic of congruencies; section 4 deals with quadratic reciprocity. Section 5 is the core of the work, dealing with the theory of binary quadratic forms over the integers, followed by a section with applications. Section 7 is on cyclotomy (division of the circle, or roots of unity).

1 January - the asteroid Ceres (named after the Roman goddess of agriculture) was first discovered by Giuseppe Piazzi. It is the largest object in the asteroid belt that lies between the orbits of Mars and Jupiter, slightly closer to Mars' orbit. Its diameter is approximately 945 kilometers.

September – Gauss gives an orbit forecast for Ceres, which was very different from others.

December - Gauss's orbit forecast for Ceres was confirmed by observations.

Gauss began to correspond with Friedrich Bessel, 14, (22 July 1784 – 17 March 1846, aged 61.6).

Germany, Freibourg im Breisgau, 510 km southwest from Brunswick, 23 March 1978, the Historical Merchants' Hall of 1520-1530, façade decorated with statues and the coat of arms of four Habsburg (1027-1780) emperors, in Freibourg im Breisgau (1120 by Duke Berthold III of Zähringen (1085-1122), elevation 278 m, population 222,000, area 153 km^2), southwest Germany, near France and Switzerland.

1805 – 9 October - Gauss, 28.4, married Johanna Osthoff, 25.4, (8 May 1780 – 11 Oct 1809, aged 29.4) for 4 years, until her premature death at only 29.4 years, after giving birth to their third child, Louis, who died in 1810.

1806 – Gauss was 29 when his first child Joseph was born (he lived 67 years, until 1873).

10 November - Gauss was 29.5 when Charles William Ferdinand, the Duke of Brunswick, died (because of his wounds from the Battle of Jena-Auerstedt (165 km southeast of Brunswick) on 14 Oct 1806, against Napoleon's marshal Davout), and the financial support from the Duke ended, after 15 years. Napoleon, 37, occupied the western German states; the Duchy of Brunswick was dissolved, and incorporated into Napoleon's Kingdom of Westphalia.

France, Lyon (43 BC), part of eastern façade of the Hôtel de Ville (1645 – 1651, 1674) de Lyon, in Place de la Comédie, across Opéra.

France, Paris, La Seine, on Parisis boat, looking upstream to the left bank, towards east: Port de Suffren with Vedettes de Paris Croisières (Cruises), near Quai Branly, the north-west and south-west sides of la Tour Eiffel (1889, 324 m, 279 m at the 3rd level observatory), with pilier nord on the left, pilier est on the center left back, pilier vest on the center front, and pilier sud on the right; the south-east end of Pont d'Iéna (1808-1814, named by Napoléon after his victory in 1806 at the Battle of Jena, 1937, 155 m by 35 m, left).

1807 – Gauss, 30, accepts a position at Göttingen (on Olbers' (49) recommendation), where he would be in charge of the new observatory, with no teaching duties.

1808 - 5 Oct - Franklin's daughter Sarah, 65, passes away. She had 8 children with Richard Bache, in 41 years of marriage (1767).

Gauss was 31 when his second child, daughter Wilhelmina (Minna), was born (she lived 38 years, until 1846).

Gauss publishes *"Theorematis arithmetici demonstratio nova"* (introduces Gauss's lemma, and uses it in the third proof of the quadratic reciprocity).

Switzerland, Geneva, the Monument (1879) for Charles II, Duke of Brunswick (30 Oct 1804 in Brunswick (Braunschweig), Germany- 19 August 1873, aged 68.8 (died in Geneva at Beau Rivage Hotel), ruled the Duchy of Brunswick 1815-1830).

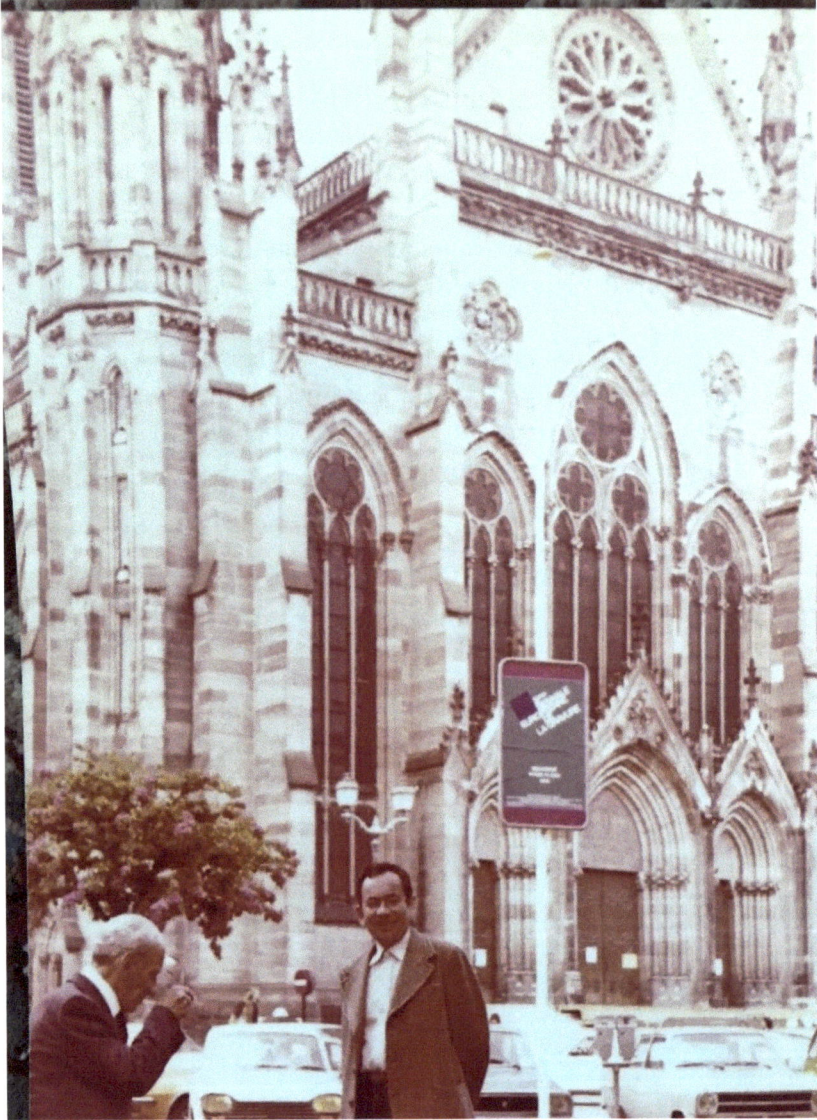

France, Mulhouse, 20 May 1978, on the main square Place de la Réunion, looking east to the southwest façade of L'église Saint-Étienne (called Cathédrale, 1866, 97 m bell tower) de Mulhouse (1150, elevation 232-338 m, population 112,000, area 22.18 km², 20 km west of Rhine River (border between France and Germany),40 km north of Switzerland, 450 km southwest of Göttingen, 1354-1515 Free Imperial City, 1515 associated with Swiss Confederation, 1798 France, 1871-1918 Germany), Alsace, east of France.

1809 - 11 Oct – Gauss was 32.4 when his wife Johanna passed away at only 29.4, after giving birth to their third child – the boy Louis, who died a few months later, in 1810. Gauss remained with two surviving children from his first marriage (Joseph, 3, and Minna, 1).

Gauss published a treatise on celestial mechanics- *Theoria Motus Corporum Coelestium in sectionibus conicis solem ambientium* (Theory of the Motion of Heavenly Bodies Moving about the Sun in Conic Sections*)*, dealing with the problem of computing the orbital parameters of a conic section from a minimum of observations, based on Kepler's laws.

1810 – 4 August - Gauss, 33.25, married Friederica Wilhelmine Waldeck, 22.29, (15 April 1788 – 12 Sep 1831, aged 43.4, daughter of a law professor at Göttingen), for 21 years, and they had 3 children: Eugen, Wilhelm and Therese Gauss (1816 – 1864, aged 48).

Gauss received Lalande Prize (named after Joseph Jérôme Lefrançois de Lalande (11 July 1732 – 4 April 1807, aged 74.7), French astronomer).

Switzerland, Fribourg (40,000 inhabitants, 550 km southwest of Göttingen), 2 May 1978, Mathematishes Institut or Département de mathématiques (the Faculty of Mathematics and Science was founded in 1896), Université de Fribourg (founded in 1889 by Georges Python (1856-1927), the only bilingual (French and German) university in Switzerland, five faculties, 10,000 students, 800 faculty), Chemin du Musée 23, close to the river La Sarine.

1811 – 29 July - Gauss was 34.25 when his 4th child Eugene was born (he lived 84.9 years, until 4 July 1896, in the USA).

Gauss published *"Summatio serierun quarundam singularium"* (determination of the sign of the quadratic Gauss sum, and uses this to give the fourth proof of the quadratic reciprocity)

1812 - Gauss, 35, starts working for 6 years, until 1818, on elliptic integrals, the arithmetic-geometric mean, and hypergeometric functions - most of it unpublished, or published posthumously.

Gauss published *Disquisitiones Generales Circa Seriem Infinitam.*

Germany, Dortmund (170 km west of Göttingen), 22 March 1978, the store Besta Hungshans (left), Avis rental service (center).

1813 – Gauss was 36 when his 5th child Wilhelm was born (he lived 66 years, until 1879).

1815 – 22 June – Gauss was 38.1 when, four days after the battle of Waterloo, Napoleon, 45.8, abdicated, and Lafayette, 57.8, arranged for the former emperor's passage to America, but the British prevented this.

1816 – Gauss was 39 when his wife Friederica (Minna), 28, gave birth to their third child (and the 6th child of Gauss) Theresa (she lived 48 years, until 1864). Now Gauss had 5 children from 2 marriages – the other four children were 10, 8, 5 and 3.

Germany, Dortmund (170 km west of Göttingen), 20 March 1978, Dortmunder Union Bier (left), Scheda (left), on a busy street only for pedestrians.

1817 – Gauss was 40 when his mother (around 60) moved in his house, and lived there for 22 years, until her death.

1818 – Gauss, 41, personally directs for 14 years, until 1832, the geodetic survey of the Kingdom of Hanover (the city of Hanover is 90 km north of Göttingen).

Gauss published *"Theorematis fundamentallis in doctrina de residuis quadraticis demonstrationes et amplicationes novae"* (he presents the fifth and sixth proofs of the quadratic reciprocity).

Germany, 23 March 1978, looking west to Neuenburg am Rhein (440 km southwest of Göttingen), near the border with France, Mullheim ahead (west), Breisach left (north), Schliengen right (south).

1821 – Gauss was 44 when his wife Friederica (Minna) began to be sick, for 10 years.

5 May - Napoleon Bonaparte passed away at 51.7, on the island of Saint Helena, at Longwood. L'Hôtel National des Invalides (1678), in Paris, the 7th arrondissement, with military museums (including details about Lafayette) and monuments, is the final burial site for Napoleon Bonaparte, 1769-1821.

Gauss published *Theoria combinationis observationum erroribus minimis obnoxiae* (the first essay concerning the calculation of probabilities as the basis of the Gaussian law of error propagation).

France, Paris, the north-west part of L'Institut de France (1795, moved in 1805 by Napoléon in this baroque building from 1684) is a revered French cultural society with five académies, the most famous being Académie française (1635) and. Académie des sciences (1666).

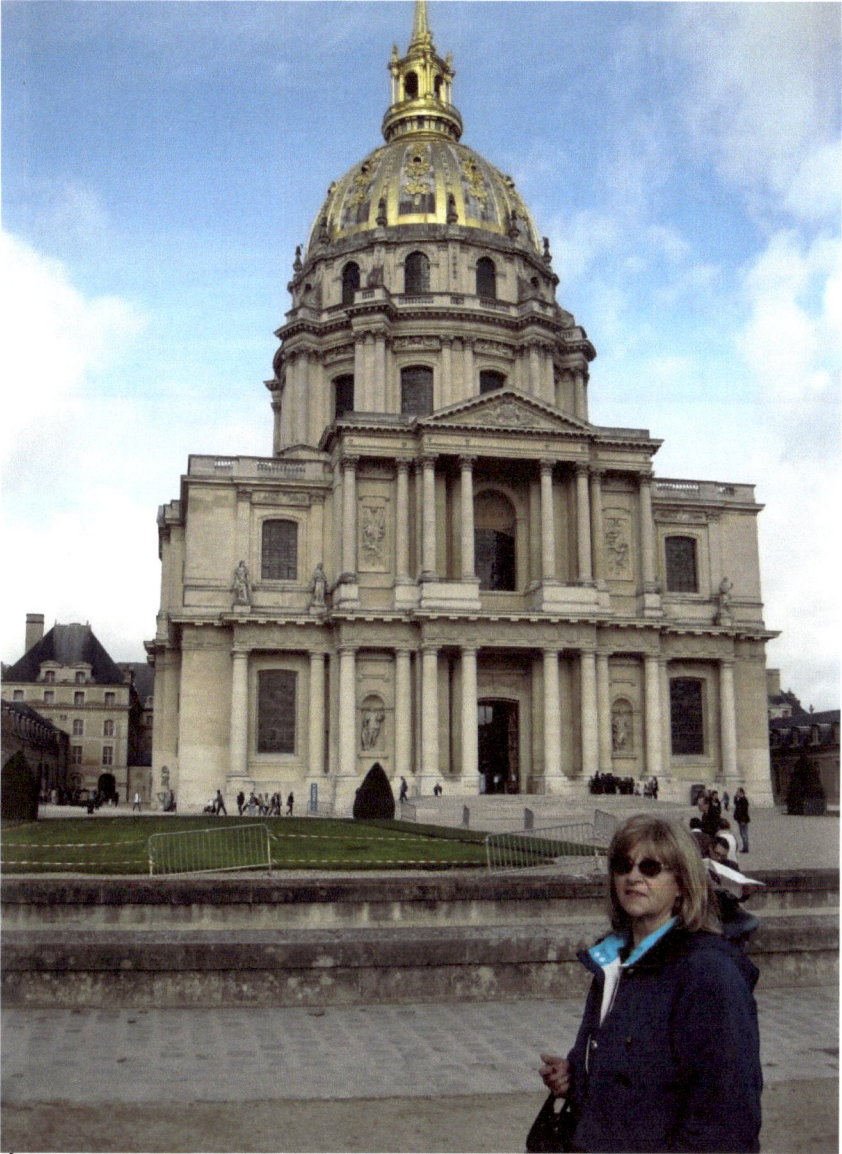

Église du Dôme (1708, 107 m height, inspired by St. Peter's Basilica in Rome, 1626) in the center of L'Hôtel National des Invalides (1678, by Louis XIV, in the 7th arrondissement, with military museums and monuments, and the burial site for Napoleon Bonaparte (1769-1821)). Napoleon was entombed under the Dôme of the Invalides, in a tomb made of red quartzite and resting on a green granite base, which was finished in 1861.

1823 – Gauss, 46, presents Copenhagen Prize Essay- general conformality criterion for mapping between two arbitrary surfaces, using the Cauchy (later Riemann contributed) equations (Baron Augustin-Louis Cauchy, now 34, (21 August 1789 – 23 May 1857, aged 67.7), French mathematician).

Gauss published *Theoria combinationis observationum erroribus minimis obnoxiae* (the second essay concerning the calculation of probabilities as the basis of the Gaussian law of error propagation).

1826 – Gauss, 49, published *Theoria combinationis observationum erroribus minimis obnoxiae* (the third essay concerning the calculation of probabilities as the basis of the Gaussian law of error propagation).

Germany, Dortmund (170 km west of Göttingen), 20 March 1978, Pezzer (center), Pelze (left).

1827 – Gauss, 50, published *Disquisitiones generales circa superficies curvas* ("General Investigations of Curved Surfaces"), which is his main work on differential geometry. He introduced the `Gaussian curvature`, and the program of investigating intrinsic differential geometry.

1828 – Gauss, 51, published in theoretical geodesy, *Determination of the latitude difference between Göttingen and Altona*, (now Altona, 225 km north of Göttingen, is the westernmost urban borough of the German city state of Hamburg, on the right bank of the Elbe river), which includes a development of the theory of least-squares fitting of data (linear regression).

Gauss published *"Theoria residuorum biquadraticorum, Commentatio prima",* a paper on the law of biquadratic reciprocity: facts about biquadratic residues, proves one of the supplements of the law of biquadratic reciprocity (the biquadratic character of 2).

Germany, Dortmund (170 km west of Göttingen), 21 March 1978, Die Stoberecke (center right).

1829 – Gauss, 52, published two short papers in mathematical physics: one on a `principle of least constraint' in mechanics, another on the equilibrium shape of a fluid touching the walls of a container (capillarity).

1830 – Gauss was 53 when his daughter Wilhelmina, 22, married Heirich Ewald, 27, (16 Nov 1803 – 4 May 1875, aged 71.5, orientalist, theologian, philosopher).

Japan, Tsukuba, 20 Nov 2008, inside the main research building, photographs with celebrity physicists, like Isaac Newton (right down), Galileo Galilei (1564 – 1642, who used the Leaning Tower of Pisa, Italy (1173 – 1372), center-right), at the High Energy Accelerator Research Organization (KEK, 1997) in Tsukuba Science City (1962), in Ibaraki Prefecture, 60 km north-east of Tokyo.

14 Aug 1977, Amsterdam (1275, population 1.3 M, elevation minus 2 m (2 m under the Atlantic Ocean), 360 km west of Göttingen): the east façade of the Royal Palace (1655 (initially Town Hall, inspired by Roman palaces, first king Louis Napoleon (1778-1846, King of Holland 1806-1810, a younger brother of Napoleon)) built on 13,659 wooden piles (like in Venezia), floor area 22,031 m^2, with yellowish sandstone from Bentheim in Germany), on the west side of Dam Square.

1831 – 12 Sep - Gauss was 54.4 when his wife Friederica (Minna) passed away at only 43.4, after 10 years of sickness. His sons Eugen, 20, and Wilhelm, 18, will emigrate to North America. The daughter Therese, 15, stayed with Gauss, and kept house for him for 23 years, until his death, when she was 38, and then she married.

Wilhelm Weber, 27, (24 Oct 1804 – 23 June 1891, aged 86.6) joins the Göttingen faculty as professor of physics; this marks the beginning of Gauss's systematic research in physics, especially potential theory.

Germany, Dortmund (170 km west of Göttingen), 16 August 1977, Willkommen in Dortmund, the south façade of the Hauptbahnhof (Railway Station).

1832 – Gauss, 55, published *"Theoria residuorum biquadraticorum, Commentatio secunda",* a second paper on the law of biquadratic reciprocity: introduces the Gaussian integers, states (without proof) the law of biquadratic reciprocity, proves the supplementary law for $1 + i$).

Gauss starts research on terrestrial magnetism, encouraged by Weber, 28, and Alexander von Humboldt, 63, (14 Sep 1769 – 6 May 1859, aged 89.6, naturalist). The goal was to map out the magnetic field of the earth, including local and temporal variations.

Gauss publishes the first foundational paper on geomagnetism.

His son Eugene, 21, who shared a good measure of his father's talent in languages, emigrated to the United States, where he was quite successful. While working for the American Fur Company in the Midwest, he learned the Sioux language. Later, he moved to Missouri, and became a successful businessman. Eugen had 8 children, one of them being Robert.

USA, Newport (1639), the west site of the Elms, 1899 - 1901, by Edward Julius Berwind (1848 – 1936), inspired from Château d'Asnières (1753) in Asnières-sur-Seine (1158, 7.9 km northwest of the center of Paris, France).

1833 – Gauss, 56, participates in the establishment of a magnetic observatory.

Gauss and Weber, 29, developed the first functioning electromagnetic telegraph, which was built in Göttingen, connecting the Astronomical Observatory (Gauss) and Weber's lab (distance: 1.2 km).

USA, Newport (1639), the northeast site of the historic Hunter House (1748, by Jonathan Nichols (1712-1756)), Georgian Colonial, Goat Island (right back).

1837 – 20 June - Gauss was 60.1 when William IV (21 August 1765 – 20 June 1837, aged 71.8, King of the United Kingdom and King of Hanover (26 June 1830 – 20 June 1837, just 6 days before 7 years), he was the third son of King George III, and succeeded his elder brother George IV), the last House of Hanover king of the UK, died, and the crown passes to Victoria, 18 years and 27 days, (24 May 1819 – 22 Jan 1901, aged 81.6, Queen 20 June 1837 – 22 Jan 1901 (63.58 years), ending the dynastic link, for 123 years, from 1714, between Hanover and England. Ernst Augustus, 66, (5 June 1771 in Buckingham House, London – 18 Nov 1851 in Hanover, aged 80.45, King of Hanover 20 June 1837 – 18 Nov 1851 (14.4 years), son of King George III of the UK) becomes King of Hanover, and sets aside the liberal constitution granted four years earlier, in 1829. Seven Göttingen professors, including the brothers Grimm (Jacob Grimm, 52, (4 Jan 1785 – 20 Sep 1863, aged 78.7, and Wilhelm Grimm, 51, (24 Feb 1786 – 16 Dec 1859, aged 73.8, philologists), Weber, 33, and Heinrich Ewald, 34, (Gauss's son-in-law from 1830, when he married Wilhelmina), refuse to sign an oath of allegiance to the new constitution, and are dismissed next year. Weber quickly gets a position in Leipzig; but this was a serious loss for Gauss.

His son Wilhelm, 24, moved to America, and settled in Missouri, starting as a farmer, and later becoming wealthy in the shoe business in St. Louis. He had 8 children.

USA, Newport, Cliff Walk (1985, 5.6 km, public access walkway that borders the Atlantic shore line, looking east at the Forty Steps, and to the Easton Bay). This is at the east end of Narragansett Avenue. On this Avenue, 800 m west, there is Osgood-Pell House, 1888, (William H. Osgood (1830-1896, zinc business)), from 1992 office for The Preservation Society of Newport County. Ochre Court (1892, 50 rooms) is 300 m southwest, on Webster Street.

UK, London, Parliament Square, the bronze statue (1973) of Sir Winston Churchill (1874-1965, Prime Minister 1940-1945, 1951-1955 (as Prime Minister he lived at 10 Downing Street, just 400 m northwest (right) from this place; as Churchill's youngest daughter, Mary Soames (1922-2014) had the run of 10 Downing Street, and helped arrange dinner with Stalin (1878-1953) in Moscow, 1942), created by the British sculptor Ivor Roberts-Jones (1913-1996).

1838 – Gauss, 61, received the Copley Medal from the Royal Society in London. In 1753 Benjamin Franklin received also the Copley Medal.

1839 – Gauss was 62 when his mother passed away at around 82.

Gauss published a second foundational paper on geomagnetism.

He taught himself Russian.

1840 – Gauss, 63, published the third foundational paper on geomagnetism. Also Gauss and Weber compiled the material for the publication of the first geomagnetic atlas.

Gauss published *Dioptrische Untersuchungen*, in which he gave the first systematic analysis on the formation of images under a paraxial approximation (Gaussian optics). Among his results, Gauss showed that, under a paraxial approximation, an optical system can be characterized by its cardinal points, and he derived the Gaussian lens formula.

USA, Newport, Chateau-sur-Mer, 1851, 17 acres, by William Shepard Wetmore (1801-1862, a merchant in the China trade).

Lausanne (Roman 150, 147,000, 41 km^2, 500 m elevation, 62 km northeast of Geneva, 610 km southwest of Göttingen, the home of the International Olympic Committee), Place de la Navigation, Château d'Ouchy (1170 by the Bishop of Lausanne, 1464 rebuilt by the Bishop of Lausanne Guillaume de Varax (1390-1466), in 1609 abandoned, in 1885 Jean-Jacques Mercier bought the land and rebuilt the castle in a neo-gothic style, between 1889 and 1893, and converted it into a luxurious hotel, which is still working).

1843 – Gauss, 66, published in theoretical geodesy *Untersuchungen über Gegenstände der Höheren Geodäsie. Erste Abhandlung* (Investigations on the foundation of higher geodesy. First treatise), based on the conformal mapping of an ellipsoid to a sphere. The geodesy work also rekindles an interest in the foundations of the geometry, from a practical standpoint (trigonometric calculations if physical space is not Euclidean, but, for example, hyperbolic).

1845 – Gauss, 68, became associated member of the Royal Institute of the Netherlands (which became the Royal Netherlands Academy of Arts and Sciences in 1851).

Netherlands, Amsterdam, 14 August 1977, in the North See harbor two big ships: Lübeck Linie (center), Naaskerk Antwerpen (right).

1846 – Gauss was 69 when his daughter Wilhelmina (Minna), from his first marriage, died at only 38.

Gauss published *Untersuchungen über Gegenstände der Höheren Geodäsie. Zweite Abhandlung* (Investigations on the foundation of higher geodesy. Second treatise),

1851 – Gauss, 74, joined as a foreign member the Royal Netherlands Academy of Arts and Sciences

Netherlands, 14 Aug 1977, Amsterdam (1275, population 1.3 M, elevation minus 2 m (2 m under the Atlantic Ocean level)): Zijkanaal G, with a bridge for the street s150, and Havenstraat on the left.

1854 – Gauss, 77, selected the topic for Bernhard Riemann's, 28, (17 Sep 1826 in Jameln, Germany – 20 July 1866 in Verbania, Italy, aged 39.8) Habilitationvortrag, *Über die Hypothesen, welche der Geometrie zu Grunde liegen.* Using this topic, Riemann delivered his inaugural lecture at Göttingen, on the hypotheses that lie at the foundations of geometry. On the way home from Riemann's lecture, Weber, 50, reported that Gauss was full of praise and excitement.

USA, Newport, Osgood-Pell House, 1888, (William H. Osgood (1830-1896, zinc business)), from 1992 office for The Preservation Society of Newport County.

Geneva (121 BC under Romans, 375 m elevation, population 200,000, area 16 km^2, 70 km northwest of Mont Blanc (4810 m), 660 km southwest of Göttingen), on Rue de la Servette (to the right, going southeast, near Rue Jean Robert Chouet ((1642-1731, physician and politician) (the street is to the left, going northeast)), a nice building having down the restaurant Le Portail Chez Rui (yellow), 1.6 km northwest from Jet d'Eau, 1.6 km southwest from Palais des Nations (UN), 1.4 km northwest from the Université de Genève (1559, John Calvin (1509-1564, aged 55)).

1855 -- 23 Feb – Gauss passes away at 77 years 9 months and 23 days, of a heart attack, in Göttingen, Kingdom of Hanover (now Lower Saxony, Germany). He is interred in the Albani Cemetery in Göttingen. Two people gave eulogies at his funeral: Gauss's son-in-law Heinrich Ewald, 52, and Wolfgang Sartorius von Waltershausen, 45.1, (17 Dec 1809 – 16 March 1876, aged 66.25, geologist), who represented the faculty in Göttingen, and was Gauss's close friend and biographer. His brain has been incorporated in the anatomical collection of the University of Göttingen. Gauss's brain was preserved, and was studied by Rudolf Wagner, 49.5, (30 July 1805 – 13 May 1864, aged 58.7, anatomist), who found its mass to be slightly above average, at 1,492 grams, and the cerebral area equal to 219,588 square millimeters. Highly developed convolutions were also found, which were suggested as the explanation of his genius.

Gauss made significant contributions to many fields, including number theory, algebra, statistics, analysis, differential geometry, geodesy, geophysics, mechanics, electrostatics, magnetic fields, astronomy, matrix theory, and optics.

Gauss' doctoral students included Johann Listing, Christian Ludwig Gerling, Richard Dedekind, Bernhard Riemann, Christian Peters, and Moritz Cantor.

Other notable students included Johann Encke, Christoph Gudermann, Peter Gustav Lejeune Dirichlet, Gotthold Eisenstein, Carl Wolfgang Benjamin Goldschmidt, Gustav Kirchhoff, Ernst Kummer, August Ferdinand Möbius, L. C. Schnürlein, Julius Weisbach, and Friedrich Bessel (epistolary correspondent).

Usually referred to as the *Princeps mathematicorum* ("the foremost of mathematicians") and "the greatest mathematician since antiquity", Gauss had an exceptional influence in many fields of mathematics and science, and is ranked among history's most influential mathematicians.

Switzerland, Geneva, from Quai Gustave Ador (1845-1928, President). Jet d'Eau (1886, 1891, 1951) – a large fountain pumping lake water at 500 liters/s to 140 m, lit up at night. It is located at the point where Lac Léman empties into the Rhône River. There are two 500 kW pumps, operating at 2,400 V, consuming one megawatt of electricity. The water leaves the nozzle (10.16 cm) at a speed of 200 km/h. At any time, there are about 7,000 liters of water in the air.

1856 – Waltershausen, 47, published *Gauss zum Gedächtnis*. This biography, published one year after the death of Carl Friedrich Gauss, is regarded as Gauss's biography as Gauss wished it to be told. It is also the source of one of Gauss's most famous mathematical quotes: *Mathematics is the queen of the sciences*, and the famous story of Gauss, as a young boy, quickly finding the sum of a long string of consecutive numbers (1 to 100).

USA, Newport, The Breakers, 1893-1895, built by Cornelius Vanderbilt II (1843-1899), 70 rooms, gross area 11,644 m², living area 5,804 m² on 5 floors, house footprint an acre (4,047 m²), 9.1 m sculpted iron gates.

Switzerland, Geneva (121 BC under Romans, 375 m elevation, population 200,000, area 16 km², 70 km northwest of Mont Blanc (4810 m)), on Rue de la Servette (to the right, going southeast, Rue Jean Robert Chouet ((1642-1731, physician and politician) (the street is to the left, going northeast)), a nice building having down the restaurant Le Portail Chez Rui (yellow), 1.6 km northwest from Jet d'Eau, 1.6 km southwest from Palais des Nations (UN), 1.4 km northwest from the Université de Genève (1559, John Calvin (1509-1564)).

From Merton Street, looking southeast to the north (left) and west (right) facades of Merton College Chapel (1294, 1425, 1451, the church of Merton College (1264, the third oldest in Oxford)); there were plans to extend this church to the west (right), but the land was leased in 1517 to Bishop Richard Foxe (1448-1528), who founded Corpus Christy College (1517), next door (west) to Merton.

UK, Oxford, from Merton St. looking south to the northern façade of the main entrance of Merton College (1264). Important personalities associated with Merton College are British chemist Frederick Soddy (1877-1956, Nobel Prize in Chemistry (1921)), poet T. S. Elliot (1888 in St Louis, U. S. – 1965 in London, England, Nobel Prize in Literature (1948)), British philosopher John R. Lucas (born 1932), British mathematician Sir Andrew Wiles (born 1953, proved Fermat's (1607-1665) Last Theorem (1637) proved after 358 years).

UK, Oxford, from the Logic Ln, looking north to the High St and the south gate of the Queen's College (1341, founded by Robert de Eglesfield (1295-1349, chaplain of the Queen consort) in honor of Queen consort Philippa of Hainault (1314-1369, wife of Edward III of England (1312-1377, Reign 1327-1377, burial Westminster Abbey, they had 13 children, and their great-grandfather was King Philip III of France (1245-1285, reign 1270-1285))), University College (1249, left).

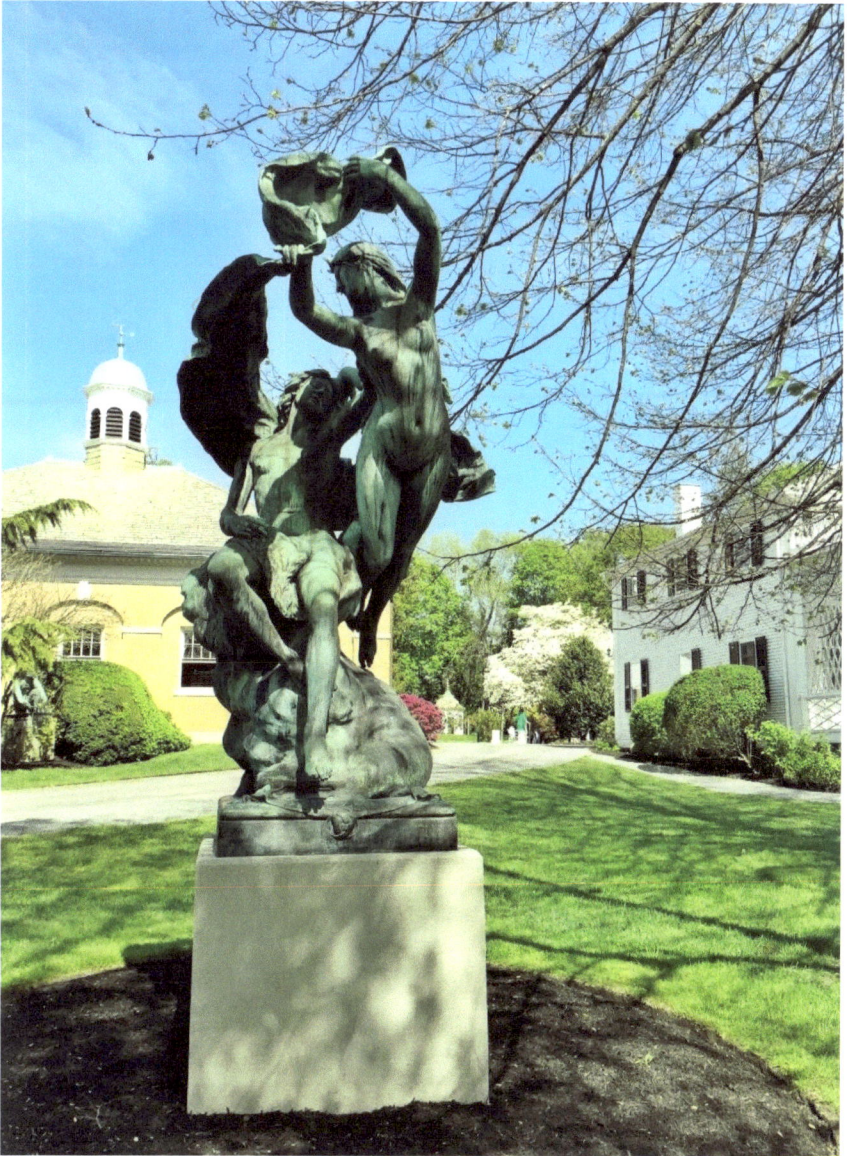

USA, Bristol, allegoric sculpture at Linden Place (1810, home of the
DeWolfs and the Colts, with Bristol Art Museum (1963, left back)
on Wardwell Street near Hope Street, located in the Ballroom of the
Linden Place.

USA, 5 Feb 2016, driving in winter 30 km northwest of Boston, on Lowell Street, in Andover (1642, incorporated 1646, population 34,000, named after Andover (circa 955, county of Hampshire, 100 km southwest of London, 60 km south of Oxford, England, population 64,000)). The highly selective Phillips Academy Andover (1778, the oldest incorporated high school in the USA) is a College preparatory high school (grades 9-12, 1122 students).

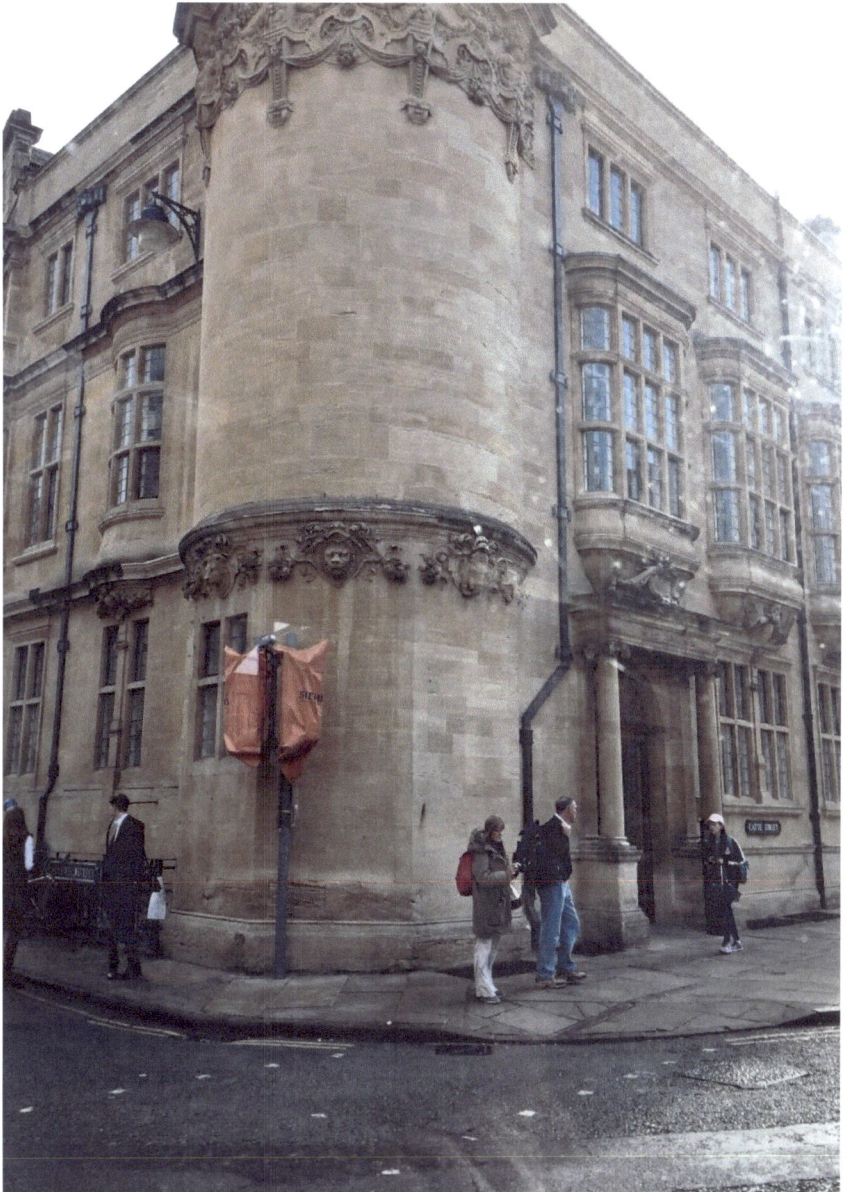

UK, Oxford, on Catte Street (to right) at Holywell St (to left), Oxford Martin School (2005, a research and policy unit of the University of Oxford, by James Martin (1933-2013, British information technology consultant and author, worked for IBM, earned a degree in physics at Keble College (1870, by John Keble (1792-1866, English churchman and poet))).

UK, Oxford, from Broad St, looking south to the northeastern side of the Sheldonian Theatre (1669, classical concerts, lectures, ceremonies, capacity 1,000, by Gilbert Sheldon (1598-1677, Archbishop of Canterbury), 3 busts (center line), Bodleian Library (1602, left back, main research library of the University of Oxford, over 12 M items).

USA, Gloucester, the northwest façade, with the main entrance, of the Hammond Castle, 1926-1929, by inventor (in the area of remote control, with over 400 patents) John Hays Hammond, Jr. (1888-1965, aged 77), as his home and laboratory, on 7 acres on the Atlantic coast in the Magnolia area of Gloucester, on Hesperus Ave. This medieval style castle has architectural elements starting from 1450, and now is a museum displaying Hammond's collection of Roman, medieval, and Renaissance relics, as well as displays about his life and inventions.

USA, Newport, sculpture Venus Italica by Antonio Canova (1757-1822) on the south side of Rosecliff, 1898-1902, Hermann Oelrichs

USA, Newport, the northwest part of Rosecliff, Hermann Oelrichs (1850-1906, shipping) and wife Theresa Fair Oelrichs (1871-1926).

www.ingramcontent.com/pod-product-compliance
Lightning Source LLC
Chambersburg PA
CBHW041310210326
41599CB00003B/51